Building and Surveying Series

(continued overleaf)

List continued from previous page

Building and Surveying Series
Series Standing Order
ISBN 0–333–71692–2 hardcover
ISBN 0–333–69333–7 paperback
(outside North America only)

You can receive future titles in this series as they are published by placing a
standing order. Please contact your bookseller or, in the case of difficulty, write
to us at the address below with your name and address, the title of the series
and the ISBN quoted above.

Customer Services Department, Macmillan Distribution Ltd
Houndmills, Basingstoke, Hampshire RG21 6XS, England

Contents

Preface

This book studies the technical basis of the environment which exists in and around buildings. The main considerations are the effects of heat, light and sound within buildings; together with supplies of water and electricity. The text also deals with climatic effects, energy use, sick buildings and other topics which underline the need for an integrated approach to the study and design of environmental services.

The book is intended for students of building, engineering and surveying who are studying environmental science at a variety of levels. The principal requirements of courses for degrees, vocational diplomas and certificates, and for examinations of professional institutes are satisfied by the contents of the book. Over a decade of publication has confirmed that the book is useful at all levels, from introductory student text to professional reference.

This fourth edition of the book contains new sections which have evolved from continuing teaching experience, together with appropriate revisions to data. The text still aims to give the user a secure technical knowledge based on accurate principles and terminology.

The book assumes a minimum prior knowledge of science and mathematics and the text highlights important facts and formulas as an aid to reference and to memory. Definitions and units are expressed in forms appropriate for this level of study and provide a link to the practical technical literature of the various topics. Where a numerical approach is required, worked examples are displayed step-by-step and supported by exercises for practice.

The style of writing has been kept simple but, at the same time, it has a technical content and accuracy appropriate to this level of study. The text is illustrated by labelled drawings which are intended to help explain the text and to act as models for student sketches. All of the subjects in the book are worthy of extended studies and it is hoped that this book will be a starting point for many further investigations.

Randall McMullan

Acknowledgements

The author and publishers thank the following organisations for permission to quote from their material, as indicated in the text:

CIBSE Guide for tables 2.1, 2.5, 3.1; by courtesy of The Chartered Institution of Building Services Engineers, London.

BRE Digests for table 7.2 and figures 7.4 and 7.5; by courtesy of the Building Research Establishment.

Introduction

HOW TO USE THIS BOOK

The topics in this book have been developed from fundamentals. The introductory section and the chapters on the basic principles of heat, light and sound will be of particular importance if you have not studied much physics before. These parts of the book will also be useful to anyone needing revision.

For the understanding of a new topic you should read the text carefully, as it has been written in a concise form. Understanding the material in this book will also help you make direct use of technical sources. A reading list is suggested at the end of the book.

For quick revision and reference some information, such as definitions and formulas, has been highlighted in the text. Other information has been presented in lists which are intended to summarise the topics and to aid the memory. The items in a list should be regarded as starting points for more comprehensive discussions of the subjects.

You need calculations in order to understand some topics, and also for passing examinations. The text emphasises those formulas which are especially useful and which may also need to be memorised. Important types of calculation are explained by carefully worked examples, using relatively simple calculations. Where further practice is relevant there are exercises at the end of the chapter.

The aim of this book, and of the associated courses and examinations, is to develop an understanding of the principles of environmental science. The content of the text itself gives an indication of the depth of knowledge normally expected at this level of study. The style of writing has been kept simple but it uses correct terminology and units. As such it may act as an example for the type of response expected when you need to display a knowledge of a topic.

The diagrams are intended to help to explain the subjects in the book. The drawings have been kept relatively simple so that they can form the basis for sketches. You should remember that accurate labels are as important as the drawings.

SI UNITS

The result of measuring a physical quantity is expressed as a number, followed by a unit. For example, length AB = 15 metres. In general:

> Physical quantity = Number × Unit

The number expresses the ratio of the measured quantity to some agreed standard or unit. Different systems of units have arisen over the years, including Imperial units and metric units. A rational and coherent version of the metric system has been developed, called the Système Internationale d'Unités, or SI.

SI units are intended for worldwide scientific, technical, and legal use. The units in this book are given in SI and reference to older units is made only where such units still linger in technical practice.

There are seven base units in the SI system, two supplementary units, and numerous derived units some of which are listed in the table of units. Derived units can be formed by combinations of base units; for example, the square metre. Some derived units are given new names; for example, the

Table I.1 *SI units*

Quantity	Symbol	SI unit	Symbol
Base units			
length	l	metre	m
mass	m	kilogram	kg
time	t	second	s
electric current	I	ampere	A
thermodynamic temperature	T	kelvin	K
luminous intensity	I	candela	cd
amount of substance		mole	mol
Supplementary units			
plane angle	θ	radian	rad
solid angle	Ω	steradian	sr
Some derived units			
area	A	square metre	m^2
volume	V	cubic metre	m^3
density	ϱ	kilogram per cubic metre	kg/m^3
velocity	v	metre per second	m/s
force	F	newton	N ($kg\,m/s^2$)
energy	E	joule	J ($N\,m$)
power	P	watt	W (J/s)
pressure	p	pascal	Pa (N/m^2)

newton is a combination of the kilogram, the metre, and the second; the pascal is a combination of the newton and the square metre.

The symbols for SI units do not have plural form and are not followed by a full stop, except at the end of a sentence. The symbols for derived units may be written in index form or with a solidus (/). For example, m s^{-2} or m/s^2.

SI prefixes

Multiplication factors are used to express large or small values of a unit. These multiples or sub-multiples are shown by a standard set of prefix names and symbols which can be placed before any SI unit.

The standard prefixes and symbols can be placed before any SI unit, with the exception of the kilogram. Multiples should be chosen so that the numerical value is expressed as a number between 0.1 and 1000.

Table I.2 *SI prefixes*

Prefix	Symbol	Multiplication factor	
tera	T	10^{12}	= 1 000 000 000 000
giga	G	10^{9}	= 1 000 000 000
mega	M	10^{6}	= 1 000 000
kilo	k	10^{3}	= 1 000
milli	m	10^{-3}	= 0.001
micro	μ	10^{-6}	= 0.000 001
nano	n	10^{-9}	= 0.000 000 001
pico	p	10^{-12}	= 0.000 000 000 0001

The Greek alphabet

The symbols for some quantities and units are taken from the Greek alphabet.

A	α	Alpha	N	ν	Nu
B	β	Beta	Ξ	ξ	Xi
Γ	γ	Gamma	O	o	Omicron
Δ	δ	Delta	Π	π	Pi
E	ε	Epsilon	P	ϱ	Rho
Z	ζ	Zeta	Σ	σ	Sigma
H	η	Eta	T	τ	Tau
Θ	θ	Theta	Y	υ	Upsilon
I	ι	Iota	Φ	ϕ	Phi
K	$\varkappa$	Kappa	X	χ	Chi
Λ	λ	Lambda	Ψ	ψ	Psi
M	μ	Mu	Ω	ω	Omega

Symbols and formulas

Some common symbols and formulas used in technical and mathematical expressions are given below.

Symbols	Meaning
Σ	sum of
$>$	greater than
$<$	less than
a^n	a raised to the power n
$\sqrt{a}$ or $a^{0.5}$	square root of a
$\log x$	common logarithm of x
π	'pi' $= 3.141\ 593$ approx.

1 Principles of Heat

A good thermal environment is a major aspect in the successful performance of a building. Both human beings and their buildings interact with the heat that surrounds them and they also contribute to this heat.

Topics that are relevant to thermal design include the requirements of human comfort, the types of heat loss and heat gain by buildings, and the nature of moisture in the air. Before these topics are considered, this chapter describes the basic nature of heat, its measurement, and its effects. The properties of gases and their effects, such as refrigeration, are also studied.

NATURE OF HEAT

Heat energy

The modern definition of heat that follows is a simple statement, but the truth of the statement was not obvious in the past and confused ideas about the nature of heat are still common.

- **_HEAT_ (_H_ or _Q_) is a form of energy**

  UNIT: joule (J)

The joule is the standard SI unit of energy as used for measuring any other form of energy. Other units of energy still found in use include the following units:

- calorie, where $1\,cal = 4.187\,J$
- kilowatt hour, where $1\,kWh = 3.6\,MJ$
- British Thermal Unit, where $1\,BTU = 1.055\,kJ$.

Heat energy is an internal molecular property of a material. Other forms of energy include mechanical energy, electrical energy, and chemical energy. These other forms of energy can all be converted to thermal energy. For example, the mechanical energy of moving surfaces is converted to heat by friction; electric currents flowing in conductors produce heat; and combustion (burning) converts the chemical energy contained in materials to heat.

Thermal energy often forms an intermediate stage in the production of other forms of energy. Most electrical energy, for example, is produced by means of the thermal energy released in the combustion of fuels. The thermal energy radiated from the Sun is also the origin of most energy used on Earth including the fossil fuels, such as coal and oil, which were originally forests grown in sunlight.

Power

Power is a measure of the rate at which work is done, or at which energy is converted from one form to another.

$$\textbf{\textit{Power}}\,(\textbf{\textit{P}}) = \frac{\text{Heat energy}\,(H)}{\text{time}\,(t)}$$

UNIT: watt (W)

By definition, 1 watt = 1 joule/second. The watt is often used in the measurement of thermal properties and it is useful to remember that it already contains information about time and there is no need to divide by seconds.

Temperature

Temperature is *not* the same thing as heat. A red-hot spark, for example, is at a much higher temperature than a pot of boiling water; yet the water has a much higher heat 'content' than the spark and is more damaging.

- **TEMPERATURE is the condition of a body that determines whether heat shall flow from it**

    UNIT: degree Kelvin (K)

See also the definitions of temperature scales.

Heat flows from objects at high temperature to objects at low temperature. When there is no net heat transfer between two objects they are at the same temperature.

Thermometers

The human body is sensitive to temperature but it is unreliable for measuring temperature. The brain tends to judge temperature by the rate of heat flow in or out of the skin. So for example, a metal surface always 'feels' colder than a plastic surface even though a thermometer may show them to be at the same temperature.

A thermometer is an instrument that measures temperature by making use of some property of a material that changes in a regular manner with changes in temperature. Properties available for such use include changes in size, changes in electrical properties such as resistance, and changes in light emissions. Some of the more common types of thermometer are described below.

Mercury-in-glass thermometers
Mercury-in-glass thermometers use the expansion of the liquid metal mercury inside a narrow glass tube. The mercury responds quickly to changes in temperature and can be used between $-39\,°C$ and $357\,°C$, which is the range between the freezing point and the boiling point of mercury.

Alcohol-in-glass thermometers
Alcohol-in-glass thermometers use coloured alcohol as the liquid in the glass tube. Alcohol expands more than mercury and can be used between $-112\,°C$ and $78\,°C$, which is the range between the freezing point and boiling point of alcohol.

Thermoelectric thermometers
Thermoelectric thermometers use the electric current generated in a thermo-couple, which is made by joining two different metals such as iron and constantan alloy. The current quickly varies with temperature and can be incorporated in remote or automatic control systems.

Resistance thermometers
Resistance thermometers use the change in electrical resistance which occurs when a metal changes temperature. Pure platinum is commonly used and the changes in its resistance can be measured very accurately by including the thermometer in an electrical circuit.

Optical pyrometers
Optical thermometers measure high temperature by examining the bright-ness and colour of the light emitted from objects at high temperatures. The light varies with temperature and is compared with a light from a filament at a known temperature.

Temperature scales

In order to provide a thermometer with a scale of numbers, two easily obtainable temperatures are chosen as upper and lower *fixed points*. The interval between these two points on the thermometer is then divided into equal parts, called degrees. The properties of water are used to define two common fixed points – the temperature at which ice just melts and the temperature of steam from boiling water – where both are measured at normal atmospheric pressure.

Celsius scale

The Celsius temperature scale numbers the temperature of the melting point of ice as 0, and the boiling point of water as 100.

- ***CELSIUS TEMPERATURE (θ) is a point on a temperature scale defined by reference to the melting point of ice and the boiling point of water***

  UNIT: degree Celsius (°C)

Degrees Celsius are also used to indicate the magnitude of a particular change in temperature, such as an increase of 20 °C. The less correct term 'centigrade' is also found in use.

Thermodynamic scale

Considerations of energy content and measurement of the expansion of gases lead to the concept of an absolute zero of temperature. This is a temperature at which no more internal energy can be extracted from a body and it occurs at -273.16 °C. The absolute (or thermodynamic) temperature scale therefore numbers this temperature as zero.

The other fixed point for the thermodynamic scale is the triple point of water; the temperature at which ice, water, and water vapour are in equilibrium (0.01 °C).

- ***THERMODYNAMIC TEMPERATURE (T) is a point on a temperature scale defined by reference to absolute zero and to the triple point of water***

  UNIT: degree Kelvin (K)

The degree Kelvin is the formal SI unit of temperature but the degree Celsius is also used in common practice. The interval of a degree Kelvin is the same size as a degree Celsius, therefore a change in temperature of 1 K is the same as a change in temperature of 1 °C.

The general relationship between the two temperature scales is given by the following formula.

$$T = \theta + 273$$

where T = Thermodynamic temperature (K)
θ = Celsius temperature (°C).

Heat capacity

The same mass of different materials can 'hold' different quantities of heat. Hence water must be supplied with more heat than the oil in order to produce the same rise in temperature. Water has a greater heat capacity than oil; a property that is not to be confused with other thermal properties such as conductivity.

The heat capacity of a particular material is measured by a value of specific heat capacity, and table 1.1 gives values for a variety of materials.

- **The *SPECIFIC HEAT CAPACITY* (*c*) of a material is the quantity of heat energy required to raise the temperature of 1 kg of that material by 1 degree Kelvin (or 1 degree Celsius)**

UNIT: J/kg K (or J/kg °C)

Table 1.1 *Specific heat capacities*

Material	Specific heat capacity (J/kg K)
Water	4190
Concrete and brickwork	3300
Ice	2100
Paraffin oil	2100
Wood	1700
Aluminium	910
Marble	880
Glass	700
Steel	450
Copper	390

Note: The values for particular building materials vary.

The heat capacity of water is higher than the heat capacities of most other substances, so water is a good medium for storing heat. The temperatures on the planet Earth are stabilised by the huge quantities of heat energy stored in the oceans and the presence of this water around islands, such as the British Isles, prevents seasonal extremes of temperature. In summer the water absorbs heat and helps to prevent air temperatures rising; in winter the heat stored in the water is available to help prevent temperatures falling. Heat exchange devices, such as boilers and heating pipes, also make use of the high heat capacity of water for transferring heat from one place to another.

Density

The heat capacities of different materials are compared on the basis of equal masses. However, the same mass of different materials may occupy different volumes of space, depending upon their densities.

Density $\left(\varrho\right) = \dfrac{\text{Mass}\ (m)}{\text{Volume}\ (V)}$

UNIT: kilogram per cubic metre (kg/m^3)

Heavyweight masonry materials, such as brick, concrete, and stone, have high densities. This means that relatively small volumes of these materials have a large mass and therefore provide a relatively high heat capacity within a small volume. An electric storage heater, for example, contains bricks which are heated by cheap rate electricity and then hold this heat for use later in the day. The heat storage provided by the brick, concrete, and stone used in construction is particularly relevant to the thermal behaviour of buildings, as discussed in later chapters.

Change of state

All matter is made from small particles called *atoms* and for most materials the smallest particle that exists independently is a group of atoms which are combined to form a *molecule*. The spacing of the molecules in a substance and the forces between them determine the phase, or state of matter, of that substance.

In the normal ranges of temperature and pressure there are three possible states of matter and they have the basic characteristics given below.

- **Solid state:** The molecules are held together in fixed positions; the volume and shape are fixed.
- **Liquid state:** The molecules are held together but have freedom of movement; the volume is fixed but the shape is not fixed.
- **Gas state:** The molecules move rapidly and have complete freedom; the volume and shape are not fixed.

The state of a substance depends upon the conditions of temperature and pressure which act on the substance. Consider, for example, the common forms of iron, water, and oxygen. At certain temperatures a material will undergo a change of state and in this change its energy content is increased or decreased.

The **absorption** of heat by a solid or a liquid can produce the following changes of state:

$$\textbf{SOLID} \xrightarrow[\text{(Melting)}]{\text{Liquefaction}} \textbf{LIQUID} \xrightarrow[\text{(Boiling, Evaporation)}]{\text{Vaporisation}} \textbf{GAS}$$

The **release** of heat from a gas or a liquid can produce the following changes of state:

$$\textbf{GAS} \xrightarrow{\text{Condensation}} \textbf{LIQUID} \xrightarrow[\text{(Fusion)}]{\text{Solidification}} \textbf{SOLID}$$

Sensible and latent heat

In order to understand how most substances behave it is useful to consider the changes of state for water. Figure 1.1 shows the effects of supplying heat energy at a constant rate to a fixed mass of ice.

Sensible heat
When the sample exists entirely in a single state of ice, water, or steam the temperature rises uniformly as heat is supplied. This heat is termed 'sensible' because it is apparent to the senses.

- *SENSIBLE HEAT* **is the heat energy absorbed or released from a substance during a change in temperature**

Latent heat
When the sample is changing from one state to another the temperature remains constant, even though heat is being supplied. This heat is termed 'latent' because it seems to be hidden.

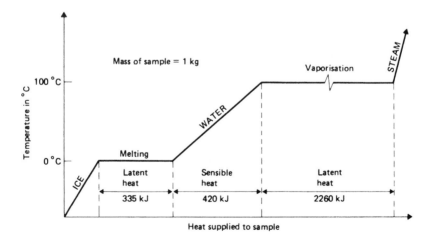

Figure 1.1 *Changes of state for water*

- *LATENT HEAT* **is the heat energy absorbed or released from a sub-stance during a change of state, with no change in temperature**

The latent heat absorbed by melting ice or by boiling water is energy which does work in overcoming the bonds between molecules. It is a less obvious, but very important, fact that this same latent heat is given back when the steam changes to water, or the water changes to ice. The latent heat changes occur for any substance and have the following general names:

$$\text{SOLID} \xrightarrow[\text{absorbed}]{\substack{\text{Latent heat} \\ \text{of fusion}}} \text{LIQUID} \xrightarrow[\text{absorbed}]{\substack{\text{Latent heat of} \\ \text{vaporisation}}} \text{GAS}$$

$$\text{GAS} \xrightarrow[\text{released}]{\substack{\text{Latent heat of} \\ \text{vaporisation}}} \text{LIQUID} \xrightarrow[\text{released}]{\substack{\text{Latent heat} \\ \text{of fusion}}} \text{SOLID}$$

A liquid may change to a gas without heat being supplied, by evaporation for example. The latent heat required for this change is then taken from the surroundings and produces an important cooling effect.

Enthalpy

Enthalpy can be described as the total heat content of a sample, with reference to 0 °C. For the particular example of water shown in figure 1.1, the steam at 100 °C has a much higher total heat content than liquid water at 100 °C. Steam at high temperature and pressure has a very high enthalpy, which makes it useful for transferring large amounts of energy such as from a boiler to a turbine. This steam is also very dangerous if it escapes.

Calculation of heat quantities

Both sensible and latent heat are forms of heat energy that are measured in joules, although they are calculated in different ways.

Sensible heat

When a substance changes temperature the amount of sensible heat ab-sorbed or released is given by the following formula:

$$H = m\,c\,\Delta\theta$$

where H = quantity of sensible heat (J)
        m = mass of substance (kg)

c = specific heat capacity of that substance (J/kg K)
$\Delta\theta = \theta_2 - \theta_1$ = temperature change (°C).

Worked example 1.1

A storage heater contains concrete blocks with total dimensions of 800 mm by 500 mm by 220 mm. The concrete has a density of 2400 kg/m³ and a specific heat capacity of 3300 J/kg K. Ignoring heat losses, calculate the quantity of heat required to raise the temperature of the blocks from 15 °C to 35 °C.

Know   Volume = $0.8 \times 0.5 \times 0.22 = 0.088\,\text{m}^3$

$$\text{Density} = \frac{\text{mass}}{\text{volume}}$$

Therefore

$$\text{mass} = \text{density} \times \text{volume}$$
$$m = 2400 \times 0.088 = 211.2\,\text{kg}$$

Using $H = m\,c\,(\theta_2 - \theta_1)$
$$= 211.2 \times 3300 \times (35 - 15)$$
$$= 13\,939\,200\,\text{J}$$

So quantity = **13.94 MJ**

Latent Heat

During a change of state in a substance the amount of latent heat absorbed or released is given by the following formula.

$$\boxed{H = m\,l}$$

where H = quantity of latent heat (J)
m = mass of substance (kg)
l = specific latent heat for that change of state (J/kg).

- **SPECIFIC LATENT HEAT (l) is a measure of the latent heat absorbed or released from a particular material for a given change of state**

  UNIT: J/kg

Specific latent heat is sometimes termed *specific enthalpy change* and some common values are as follows:

Specific latent heat of ice = 335 000 J/kg
= 335 kJ/kg

Specific latent heat of steam $= 2\,260\,000\,$J/kg
$$= 2260\,\text{kJ/kg}$$

Worked example 1.2

Calculate the total heat energy required to completely convert 2 kg of ice at 0 °C to steam at 100 °C. The specific heat capacity of water is 4190 J/kg °C. The specific latent heats are 335 kJ/kg for ice and 2260 kJ/kg for steam.

Divide the process into sensible and latent heat changes:

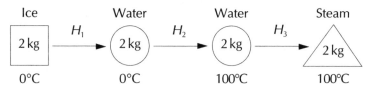

Step 1: Changing ice at 0 °C to water at 0 °C requires latent heat.

> Using $H = ml$
> $H_1 = 2 \times 335\,000 = 670\,000\,$J
> $H_1 = 670\,\text{kJ}$

Step 2: Changing water at 0 °C to water at 100 °C requires sensible heat.

> Using $H = m\,c\,\Delta\theta$
> $H_2 = 2 \times 4190 \times 100 = 838\,000\,$J
> $H_2 = 838\,\text{kJ}$

Step 3: Changing water at 100 °C to steam at 100 °C requires latent heat.

> Using $H = ml$
> $H_3 = 2 \times 2\,260\,000 = 4\,520\,000\,$J
> $H_3 = 4520\,\text{kJ}$
> Total heat $= H_1 + H_2 + H_3$
> $= 670 + 838 + 4520$
> So total heat $= \mathbf{6028\,kJ}$

Expansion

Most substances expand on heating and contract on cooling. If the natural expansion and contraction of a body are restricted then very large forces may occur. Different substances expand by different amounts and the coefficient of linear thermal expansion is a measure of the relative change of length. Superficial (area) expansion and cubical (volumetric) expansion can be predicted from the linear expansion.

Solids

The coefficient of linear expansion for steel is about $12 \times 10^{-6}/°C$, which means that a steel bar increases its length by 12/1 000 000 for each degree of temperature rise. Concrete expands at a similar rate to steel. The expansion of aluminium is about twice that of steel and the expansion of plastics is up to ten times that of steel.

Allowance must be made in constructions for the effects of expansion, particularly for concrete, metals, and plastics. The result of destructive expansion can be seen in the twisted girders of a building after a fire. Expansion effects can also be useful. Heated rivets, for example, contract on cooling and tighten the joint between metal plates. The unequal expansion of two metals deforms a bimetallic strip, which can then be used as a temperature switch like the thermostat shown in figure 1.2.

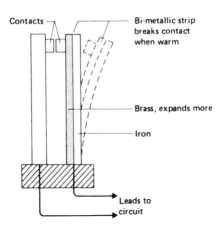

Figure 1.2 *Simple thermostat*

Liquids

Liquids tend to expand more than solids, for the same temperature rise. The expansion rates of different liquids vary and the expansion of alcohol is about five times that of water. Most liquids contract upon cooling but water is unusual in that its volume increases as it cools from 4 °C to 0 °C. At 0 °C the volume of water expands by a larger amount as it changes to the solid state of ice.

Thermometers make use of the expansion of liquids but in the hot water systems of buildings and car engine blocks the expansion is troublesome if it is not allowed for. Rainwater that freezes and expands in the pores and crevices of concrete or stonework will disrupt and split the material. This process has, over long periods of time, destroyed whole mountain ranges.

Gases

The expansion of gases is hundreds of times greater than the expansion of liquids. This will not be apparent if the gas is confined in a container because the pressure will then increase instead of the volume; this behaviour will be described in the later section on gases and vapours. If a gas is allowed to expand under conditions of constant pressure then the coefficient of volumetric expansion is found to be 1/273 per degree, starting at 0 °C.

The concept of an absolute zero of temperature at −273 °C was a result of imagining the effect of cooling an ideal gas. Starting at 0 °C, this ideal gas would shrink in size by 1/273 for each drop in temperature of 1 °C. At −273 °C the volume of the gas would therefore be zero, and matter would have disappeared. It is not possible to achieve this condition, although it is possible to approach close to it. Real gases do not actually stay in the gas state at very low temperatures, but the concept of an absolute zero of temperature remains valid.

HEAT TRANSFER

Heat energy always tends to transfer from high temperature to low temperature regions. There is no real thing as 'cold' that flows into warm places, even though the human senses may interpret the loss of heat energy as a 'cold flow'. If several bodies at different temperatures are close together then heat will be exchanged between them until they are at the same temperature.

This equalising of temperature can occur by the three basic processes of heat transfer given below:

- Conduction
- Convection
- Radiation.

Heat may also be transferred by the process of evaporation when latent heat is absorbed by a vapour in one place and released elsewhere.

Conduction

If one end of a metal bar is placed in a fire then, although no part of the bar moves, the other end will become warm. Heat energy travels through the metal by the process of conduction.

- *CONDUCTION* is the transfer of heat energy through a material without the molecules of the material changing their basic positions

Conduction can occur in solids, liquids, and gases although the speed at which it occurs will vary. At the place where a material is heated the molecules gain energy and this energy is transferred to neighbouring molecules which then become hotter. The transfer of energy may be achieved by the drift of *free electrons*, which can move from one atom to another, especially in metals. Conduction also occurs by vibrational waves called *phonons*, especially in non-metals.

Different materials conduct heat at different rates and the measurement of thermal conductivity is described in the next chapter. Metals are the best conductors of heat, because of the free electrons that they possess. Good conductors have many applications for the efficient transfer of heat, such as in boilers and heating panels.

Poor conductors are called insulators and include most liquids and gases. Porous materials that trap a lot of air tend to be good insulators and are of particular interest in controlling heat conduction through the fabric of buildings.

Convection

Air is a poor conductor of heat yet it is still possible to heat all the air in a room from a single heating panel; by the process of convection.

- *CONVECTION* is the transfer of heat energy through a material by the bodily movement of particles

Convection can occur in fluids (liquids and gases), but never in solids. Natural convection occurs when a sample of fluid, such as air, is heated

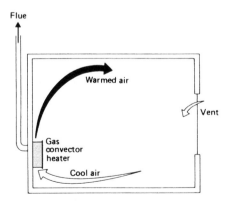

Figure 1.3 *Convection currents in room*

and so expands. The expanded air is less dense than the surrounding air and the cooler air displaces the warmer air causing it to rise. The new air is then heated and the process is repeated, giving rise to a 'convection current'.

Natural convection occurs in hot water storage tanks, which are heated by an electric element or heat exchanger coils near the bottom of the tank; convection currents ensure that all the water in the tank is heated. The term *stack effect* describes the natural convection that occurs in buildings causing warm air to flow from the lower to the upper stories. Forced convection uses a mechanical pump to achieve a faster flow of fluid, such as in the water cooling of a car engine or in a small-bore central heating system.

Radiation

Heat is transferred from the Sun to the Earth through space where conduction and convection is not possible. The process of radiation is responsible for the heat transfer through space and for many important effects on Earth.

- *RADIATION* is the transfer of heat energy by electromagnetic waves

Heat radiation occurs when the thermal energy of surface atoms in a material generates electromagnetic waves in the infra-red range of wavelengths. These waves belong to the large family of electromagnetic radiations, including light and radio waves, whose general properties are described in chapter 5.

The rate at which a body emits or absorbs radiant heat depends upon the nature and temperature of its surface. Rough surfaces present a larger total area and absorb or emit more heat than polished surfaces. Surfaces which appear dark, because they absorb most light, also absorb most heat. Good absorbers are also good emitters. Poor absorbers are also poor emitters. In general, the following rules are true:

- Dull black surfaces have the **highest** absorption and emission of radiant heat
- Shiny silver surfaces have the **lowest** absorption and emission of radiant heat.

These surface properties of radiation are put to use for encouraging heat radiation as, for example, in a blackened solar energy collector. The surface properties are also used for discouraging heat radiation as, for example, by the use of aluminium foil insulation.

The rate at which a body emits heat increases with the temperature of the body. Every object is continuously emitting and absorbing heat to and from its surroundings. Prévost's Theory of Exchanges explains that the balance of these two processes determines whether or not the temperature of the object rises, falls, or stays the same.

The wavelengths of the radiation emitted by a body also depend upon the temperature of the body. High temperature bodies emit a larger proportion of short wavelengths, which have a better penetration than longer wavelengths. The short wavelengths emitted by hot bodies also become visible at about 500 °C when they first appear as dull red.

The greenhouse effect

The 'greenhouse effect', as illustrated in figure 1.4, is one result of the differing properties of heat radiation when it is generated by bodies at different temperatures. The high-temperature Sun emits radiation of short wavelength which can pass through the atmosphere and through glass. Inside the greenhouse or other building this heat is absorbed by objects, such as plants, which then re-radiate the heat. Because the objects inside the greenhouse are at a lower temperature than the Sun the radiated heat is of longer wavelengths which cannot penetrate glass. This re-radiated heat is therefore trapped and causes the temperature inside the greenhouse to rise.

The atmosphere surrounding the Earth also behaves as a large 'greenhouse' around our world. Changes to the gases in the atmosphere, such as increased carbon dioxide content from the burning of fossil fuels, can act like a layer of glass and reduce the quantity of heat that the planet Earth

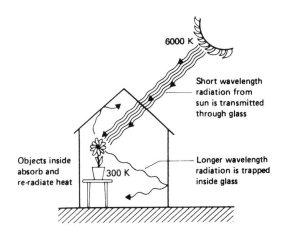

Figure 1.4 *Greenhouse effect*

would otherwise radiate back into space. This particular greenhouse effect therefore contributes to *global warming.*

GASES AND VAPOURS

Gases

The gas state is one of the three principal states in which all matter exists. According to the *Kinetic Theory,* the molecules of a gas are always in motion and their velocity increases with temperature. When the molecules are deflected by the walls of a container there will be a change in their momentum and a force imparted to the wall. The collisions of many molecules acting on a particular area will be detected as pressure.

$$\textbf{Pressure } (p) = \frac{\text{Force } (F)}{\text{Area } (A)}$$

UNIT: pascal (Pa)

By definition: 1 pascal = 1 newton/metre^2 (1 N/m^2).
Other units which may be found in use include the following:

millibars (mb), where 1 mb = 101 Pa
mm of mercury, where 1 mm = 133 Pa

Gas laws

Heating a gas increases the velocity and the kinetic energy of the molecules. If the gas is free to expand then the heated molecules will move further apart and increase the volume of gas. If the volume is fixed the heated molecules will exert a greater force at each collision with the container and so the pressure of the gas increases. The gas laws are an expression of the relationships between the temperature, volume, and pressure of a constant mass of gas.

Boyle's Law
For a fixed mass of gas at constant temperature, the volume (V) is inversely proportional to the pressure (p):

$p V$ = constant   or $p_1 V_1 = p_2 V_2$

Charles' Law
For a fixed mass of gas at constant pressure, the volume (V) is directly proportional to the thermodynamic temperature (T):

V = constant $\times T$

Pressure Law

For a fixed mass of gas at constant volume, the pressure (p) is directly proportional to the thermodynamic temperature (T):

$$p = \text{constant} \times T$$

General Gas Law

The relationships between the pressure, volume, and temperature of a gas can be combined into one expression:

$$\frac{pV}{T} = \text{constant} \qquad \text{or} \qquad \frac{p_1V_1}{T_1} = \frac{p_2V_2}{T_2}$$

Note: temperature must always be in degrees Kelvin.

Dalton's Law of Partial Pressures

Where there is a mixture of different gases, each gas exerts an individual partial pressure and has the following features:

- The partial pressure exerted by each gas is independent of the pressure of the other gases
- The total pressure of the mixture equals the sum of the partial pressures.

Standard temperature and pressure, STP

In order to compare gases measured under different conditions of temperature and pressure it is convenient to convert them to the following standard temperature and pressure (STP):

Standard temperature = 0 °C = 273 K

Standard pressure = 101.3 kPa
= 760 mm of mercury $\Big\}$ 1 atmosphere

Worked example 1.3

At 20 °C temperature and 200 kPa pressure a certain sample of gas occupies a volume of 3 litres. What volume will this sample occupy at standard temperature and pressure?

Initial conditions: $p_1 = 200$ kPa, $V_1 = 3$ litres, $T_1 = 283 + 20 = 293$ K

Final conditions: $p_2 = 101.3$ kPa, $V_2 = ?$, T_2

Using $\dfrac{p_1V_1}{T_1} = \dfrac{p_2V_2}{T_2}$

$$\frac{200 \times 3}{293} = \frac{101.3 \times V_2}{273}$$

Therefore

$$V_2 = \frac{200 \times 3 \times 273}{293 \times 101.3} = 5.519$$

So final volume = **5.519 litres**

Vapours

A vapour is a material in a special form of the gas state and has some different properties from those of a gas. For example, when a vapour is compressed the pressure increases until, at a certain point, the vapour condenses to a liquid.

- **A *VAPOUR* is a material in the gas state which can be liquefied by compression, without change in temperature**

The *critical temperature* of a substance is the temperature above which a vapour is not able to exist. Table 1.2 gives the critical temperatures of some substances relevant to heating and refrigeration. So it is seen, for example, that steam is a vapour at 100 °C but a gas at 500 °C.

The atmosphere

Air is a mixture of gases and has the following percentage composition when it is clean and dry:

Nitrogen (N_2)	78 per cent
Oxygen (O_2)	21 per cent
Carbon dioxide (CO_2) and other gases	1 per cent

Table 1.2 *Critical temperatures*

Substance	Critical temperature (°C)
Oxygen (O_2)	−119
Air	−141
Carbon dioxide (CO_2)	31
Ammonia (NH_3)	132
Water (H_2O)	374

The atmosphere is the collection of gases that surround the surface of the Earth. In addition to air the atmosphere contains up to several per cent of water vapour, and may contain local pollution products.

At the surface of the Earth the atmosphere produces a fluid pressure that depends upon the average density and the height of the atmosphere above that point. This pressure acts in all directions and varies with altitude and with local weather conditions. At sea level atmospheric pressure has a standard value of 101.3 kPa and is measured by barometers.

The simple *mercury barometer* consists of a glass tube sealed at one end which is filled with mercury and inserted upside down into a dish of mercury. A certain height of mercury remains in the tube, its weight balanced by the force resulting from atmospheric pressure. A standard atmosphere supports a 760 mm column of mercury and this height changes with changes in atmospheric pressure. The absolute unit of pressure is the pascal but it is also convenient to refer to pressure in terms of height, such as mm of mercury.

The *aneroid barometer* uses a partially evacuated box of thin metal whose sides move slightly with changes in atmospheric pressure. The movements are magnified by levers and shown by a pointer on a scale. Aneroid barometers are commonly used as altimeters for aircraft and as household weather gauges.

REFRIGERATORS AND HEAT PUMPS

A refrigeration cycle causes heat energy to be transferred from a cooler region to a warmer region. This is against the natural direction of heat flow and can only be achieved by supplying energy to the cycle. This movement of heat can be used for cooling, as in a refrigerator, or for heating, as in a heat pump. A refrigeration cycle can also be the basis of a single plant capable of either heating or cooling a building, according to needs.

The refrigeration cycle

A refrigeration system operates by absorbing heat, transferring that heat elsewhere and then releasing it. The most common forms of refrigeration make use of the following two physical mechanisms that have already been described:

- Latent heat changes that occur with changes of state
- The behaviour of vapours when compressed and expanded.

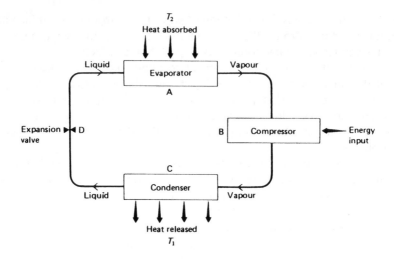

Figure 1.5 *Compression refrigerator and heat pump cycle*

A refrigeration system employs a volatile liquid called a refrigerant which undergoes the cycle described below and illustrated in figures 1.5 and 1.7.

Stage A: The liquid refrigerant is allowed to evaporate. The latent heat required for the evaporation is extracted from the surroundings of the evaporator, which acts as a cooler.

Stage B: The pressure of the refrigerant vapour is increased by means of a compressor or other mechanism. This pressurisation raises the temperature of the refrigerant and requires an input of energy into the system.

Stage C: The refrigerant vapour condenses to liquid when it cools below its boiling point. The latent heat released is emitted by the condenser which acts as a heater.

Stage D: The liquid refrigerant is passed through an expansion valve and its pressure drops. The refrigerant then evaporates and can repeat the cycle.

Refrigerants

Not every gas behaves in a way that is suitable for the cycle of vapour–liquid–vapour changes used in refrigeration. Air, for example, does *not* normally turn to liquid when it is condensed, such as inside a car tyre. A refrigerant must therefore be below its critical temperature, or it will not liquefy; and it must be above its melting point, or else it will

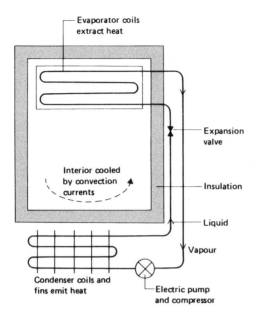

Figure 1.6 *Compression refrigerator layout*

solidify. In general a useful refrigerant should possess the following properties:

- Low boiling point
- High latent heat of vaporisation
- Easy liquefaction by compression
- Stable, non-toxic and non-corrosive
- Environmental safety.

It is possible to use air or water as refrigerants but the more common substances employed are described below.

CFCs

Chlorofluorocarbon (CFC) is a mixture of organic compounds containing carbon, chlorine, and fluorine which has been available in various commercial formulations such as *freon*. These compounds have been widely used in air conditioning and refrigeration systems, including domestic refrigerators. They had desirable properties of being colourless, non-flammable, non-toxic, and non-corrosive.

It is now accepted that CFCs help cause a reduction of the *ozone layer* in the upper atmosphere of the Earth and contribute to global warming of the World's weather systems. International Conventions have agreed to phase out the production and use of CFCs and to use safes alternatives.

CFC replacements
New compounds of hydrochlorofluorocarbon (HCFC) and hydro-fluorocarbon (HFC) are being introduced to replace CFC refrigerants. Other systems such as ammonia refrigerants and absorption cycles can also be considered for some new installations.

Ammonia
Ammonia (NH_3) is used in absorption refrigeration systems and in some industrial refrigeration and air conditioning plants. Ammonia is an efficient refrigerant but its toxicity and flammability requires special safety measures.

Carbon dioxide
Carbon dioxide (CO_2) is used in some large refrigeration plants and requires high operating pressures.

Refrigerators

Compression refrigeration cycle
The compression refrigeration cycle is driven by an electric compressor which also circulates the refrigerant. The operating principles of the compression cycle are shown schematically in figure 1.5.

A domestic refrigerator usually uses the compression cycle and a layout of a practical refrigerator is shown in figure 1.6. The evaporator coils are situated at the top of the compartment to where natural convection currents aid the flow of heat. The heat is extracted and transferred to the condenser coils which release the heat outside the refrigerator with the help of cooling fins. The temperature of the interior is regulated by a thermostat which switches the compressor motor on and off.

Absorption refrigeration cycle
The absorption refrigeration cycle, shown in figure 1.7, operates without a compressor and needs no moving parts. The principle of evaporative cooling is the same as for the compression cycle but the driving mechanism is not as easy to explain.

The refrigerant is pressurised and circulated by a generator or 'boiler', which can be heated by any source. A concentrated solution of ammonia in water is heated by the generator and the ammonia is driven off as a vapour, leaving the water. The pressure of the ammonia vapour increases and it passes through the condenser and evaporator, so acting as a refrigerant. In the absorber the ammonia is redissolved by water flowing from the generator and the cycle then repeats.

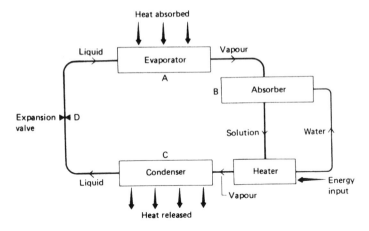

Figure 1.7 *Absorption refrigeration cycle*

The absorption cycle is used in domestic gas-operated refrigerators and for commercial plants, particularly where there is a source of waste heat which can be used to drive the cycle.

Heat pumps

Every object in the environment is a potential source of heat, even ground that is frozen solid. The quantity and the temperature of this 'low grade' heat is usually too low for it to be useful. With some sources however, it can be profitable to 'pump' its heat to a higher temperature.

- **A *HEAT PUMP* is a device which extracts heat from a low temperature source and upgrades it to a higher temperature**

The heat pump employs a standard refrigeration cycle and the compression system, shown in figure 1.5, is the commonest type:

Stage A: The evaporator coils are placed in the low temperature heat source and heat is absorbed by the evaporating refrigerant.

Stage B: The temperature of the refrigerant is increased as it is compressed, the extra energy being supplied by the electric motor.

Stage C: The refrigerant condenses and releases heat energy at a useful temperature. The condenser coils are used to heat air or some other medium.

Stage D: The refrigerant is expanded and it evaporates for a repeat of the cycle.

Heat pump efficiency

The main feature of the heat pump is that it produces more usable energy than it consumes. This efficiency is measured by a coefficient of performance for heating.

- The *Coefficient of Performance (COP$_H$)* of a heat pump is the ratio of heat output to the energy needed to operate the pump

$$COP_H = \frac{\text{Heat energy output}}{\text{Pump energy input}} = \frac{T_1}{T_1 - T_2}$$

where T_1 = absolute temperature of heat output (K)
T_2 = absolute temperature of heat source (K).

The above expressions show that the COP_H is always greater than unity, so that a heat pump always gives out more energy than is directly supplied to the pump. It therefore uses electricity more efficiently than any other system. The temperature equation also shows that the COP_H decreases as the temperature of the heat source decreases, which unfortunately means lower efficiency in colder weather.

The COP_H values achieved by heat pumps under practical conditions are between 2 and 3. These values take account of the extra energy needed to run circulatory fans and pumps, to defrost the heat extraction coils, and to supply supplementary heating if necessary. A heat pump does not actually give energy for nothing, as might be suggested by the coefficient of performance. The bottom line of the COP_H only takes account of the input energy from the user and ignores the energy taken from the source.

Heat sources for heat pumps

The extraction coils (evaporator) of a heat pump should ideally be placed in a heat source which remains at a constant temperature.

Air
Air is the most commonly used heat source for heat pumps but has the disadvantage of low thermal capacity. Outside air also has a variable temperature and when the air temperature falls the COP_H also falls. The exhaust air of a ventilation system is a useful source of heat at constant temperature.

Earth
The ground can be a useful source of low-grade heat in which to bury the extraction coils of a heat pump. Ground temperatures remain relatively

constant but large areas may need to be used, otherwise the heat pump will cause the ground to freeze.

Water
Water is a good heat source for heat pumps provided that the supply does not freeze under operating conditions. Rivers, lakes, the sea and supplies of waste water can be used.

Exercises

1 (a) Convert the following temperatures from degrees Celsius to degrees Kelvin: 20 °C; 400 °C; −10 °C.
  (b) Convert the following temperatures from degrees Kelvin to degrees Celsius: 200 K; 300 K; 773 K.

2 Explain, using accurate terminology, the reasons for the following observations:
  (a) The meat and gravy inside a hot pie burn the tongue more easily than the surrounding pastry, even though all parts of the pie are at the same temperature.
  (b) A concrete floor feels colder to walk on than carpet, even though both are at the same temperature.

3 Calculate the total heat required to convert 2 kg of water at 60 °C completely to steam at 100 °C. (Given: the specific heat capacity of water is 4200 J/kg K and the specific latent heat of steam is 2260 kJ/kg.)

4 An insulated hot water storage cylinder has internal dimensions of 0.6 m dia. and 1.5 m height. A heating element at the bottom of the cylinder supplies 74.81 MJ of heat to the water which is at an initial temperature of 8 °C. Ignore any heat losses from the water and heat gains by the cylinder itself.
  (a) Calculate the mass of water present in the cylinder.
  (b) Calculate the temperature of the water after heating.
  (c) Calculate the effective power of the heating element if the above temperature rise takes 4 hours.
  Given: density of water = 1000 kg/m^3; specific heat capacity of water = 4200 J/kg K.

5 Explain, with accurate reference to appropriate mechanisms of heat transfer, the following effects:
  (a) The ability of a vacuum flask to keep liquids either hot or cold.
  (b) The ability of silver paint on a roof to reduce heat loss from a building.

6 At standard temperature and pressure a certain sample of gas occupies 50 litres in volume. Calculate the pressure required to compress this

sample to a volume of 20 litres while allowing the temperature to rise to 30 °C.

Answers

1 (a) 293 K; 673 K; 263 K
  (b) −73 °C; 27 °C; 500 °C

3 4856 kJ

4 (a) 424.1 kg; (b) 50 °C; (c) 5195 W

6 281 kPa

2 Thermal Insulation

In order to maintain a constant temperature within a building it is necessary to restrict the rate at which heat energy is exchanged with the surroundings. Keeping heat inside a building for as long as possible conserves energy and reduces heating costs.

Thermal insulation is the major factor in reducing the loss of heat from buildings. Adequate insulation should be a feature of good initial design but insulation can also be added to existing buildings. The relatively small cost of extra insulating materials is quickly paid for by the reduction in the size of the heating plant required and by the annual savings in the amount of fuel needed. These fuel savings continue throughout the life of the building.

One of the other benefits of good thermal insulation is that the risk of surface condensation is reduced because of the warmer internal surfaces. Good insulation can also reduce the time taken for a room to heat up to a comfortable temperature; in a room that is unoccupied during the day, for example.

It is useful to remember that good thermal insulation will also reduce the flow of heat into a building, when the temperature outside is greater than the temperature inside. In other words, a well-insulated structure will, if ventilation is controlled, stay cooler in the summer than a poorly-insulated structure. In a large building this insulation will give savings in the energy needed to run the cooling plant. Some office buildings use more energy for summer cooling than for winter heating.

INSULATING MATERIALS

A thermal insulator is a material which opposes the transfer of heat between areas at different temperatures. In present-day buildings the main method of heat transfer is by conduction, but the mechanisms of convection and radiation are also relevant.

The best insulating materials have their atoms spaced well apart, so these materials will also have low density and low conductivity. Gases, which

31

have the most widely-spaced atoms, are the best insulators against *conduction*. Air, which is a mixture of gases, is the basis of insulators such as aerated lightweight concrete, expanded plastics, and cavities.

For air to act as an insulator it must be stationary, otherwise moving air is allowed to move will transfer heat by *convection*. Construction methods such as weather stripping of windows and doors, for example, prevent convection heat loss by restricting the flow of air in and out of a building.

Heat transfer by *radiation* is restricted by using surfaces that do not readily absorb or emit radiant heat. Such surfaces, which look shiny, reflect the electromagnetic waves of heat radiation. Aluminium foil, thin enough to make conduction negligible, is an example of this type of insulator.

Types of insulator

Thermal insulators used in construction are made from a wide variety of raw materials and marketed under numerous trade names. These insulation products can be grouped by form under the general headings given below:

- **Rigid preformed materials.** Example: aerated concrete blocks
- **Flexible materials.** Example: fibreglass quilts
- **Loose fill materials.** Example: expanded polystyrene granules
- **Materials formed on site.** Example: foamed polyurethane
- **Reflective materials.** Example: aluminium foil.

Properties of insulators

When choosing materials for the thermal insulation of buildings the physical properties of the material need to be considered. An aerated concrete block, for example, must be capable of carrying a load. The properties listed below are relevant to many situations, although different balances of these properties may be acceptable for different purposes:

- Thermal insulation suitable for the purpose
- Strength or rigidity suitable for the purpose
- Moisture resistance
- Fire resistance
- Resistance to pests and fungi
- Compatibility with adjacent materials
- Harmless to humans and the environment.

The measurement of thermal insulation is described in the following sections. As well as resisting the passage of moisture it is important that a material is able to regain its insulating properties after being made wet, perhaps during the construction of a building. The fire resistance of many plastic materials, such as ceiling tiles, is seriously altered by the use of some types of paints and manufacturers' instructions must be followed. Many bituminous products tend to attack plastics materials and this should be considered when installing the materials.

Thermal conductivity, *k*-value

In order to calculate heat transfer and to compare different materials it is necessary to quantify just how well a material conducts heat.

- **THERMAL CONDUCTIVITY (*k*) is a measure of the rate at which heat is conducted through a particular material under specified conditions**

    UNIT: W/m K (W/m °C may also be found in use)

This coefficient of thermal conductivity, or *k-value,* is measured as the heat flow in watts across a thickness of 1 m for a temperature difference of 1 degree (K or °C) and a surface area of $1\,m^2$.

Different techniques of measurement are needed for different types of material. Referring to figure 2.1 the general formula is

$$\frac{H}{t} = \frac{kA(\theta_1 - \theta_2)}{d}$$

where
k = thermal conductivity of that material (W/m K)
H/t = rate of heat flow between the faces (J/s = W)
A = cross-sectional area of the sample (m^2)
$\theta_1 - \theta_2$ = temperature difference between the faces (°C)
d = distance between the faces (m).

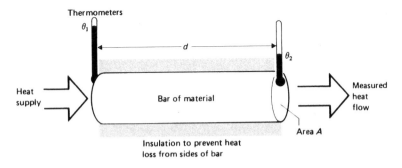

Figure 2.1 *Measurement of thermal conductivity*

Resistivity (r) is an alternative index of conduction in materials and is the reciprocal of thermal conductivity, so that $r = 1/k$. Similarly, the unit of resistivity is the reciprocal of the unit *k*-value unit: m K/W.

The thermal conductivities of some building materials are given in table 2.1. These values are a selection of measured values commonly used for standard calculations. It is important to remember that the thermal conductivity of practical building materials varies with moisture content as the presence of water increases conduction.

Variations in density have significant effects on the k-values of brickwork, concrete and stone, and more comprehensive tables of thermal conductiv-

Table 2.1 *Thermal conductivity of typical materials*

Material	k-value (W/m K)
Aluminium alloy, typical	160.00
Asbestos–cement sheet	0.40
Asphalt roofing (1700 kg/m³)	0.50
Bitumen felt layers (1700 kg/m³)	0.50
Brickwork, exposed (1700 kg/m³)	0.84
Brickwork, internal (1700 kg/m³)	0.62
Concrete, dense (2100 kg/m³)	1.40
Concrete, lightweight (1200 kg/m³)	0.38
Concrete block, medium weight (1400 kg/m³)	0.51
Concrete block, lightweight (600 kg/m³)	0.19
Copper, commercial	160.00
Corkboard	0.042
Fibre insulating board	0.050
Glass	1.022
Glass wool, mat or fibre	0.04
Hardboard, standard	0.13
Mineral wool	0.039
Plaster, dense	0.50
Plaster, lightweight	0.16
Plasterboard	0.16
Polystyrene, expanded (EPS)	0.035
Polystyrene, solid	0.17
Polyurethane (foamed) board	0.025
PVC flooring	0.040
Rendering, external	0.50
Screed (1200 kg/m³)	0.41
Steel, carbon	50.00
Stone, sandstone	1.3
Tile (1900 kg/m³)	0.84
Timber, softwood	0.13
Timber, hardwood	0.15
Urea formaldehyde foam (UF)	0.040
Woodwool slab	0.10

Note: Values are adapted from the *CIBSE Guide*. Use manufacturer's data for precise values.

ities should be used to obtain specific values. Manufacturers of insulating materials publish the *k*-values of their particular products and, when supported by recognised test certificates, these values should be used in calculations.

Emissivity and absorption

The ability of a material to absorb or give off radiant heat is a property of the surface of the material. Rough black surfaces absorb most heat and emit most heat. Conversely, shiny silvered surfaces absorb least heat and emit least heat.

To specify these properties of a surface coefficients of emission and absorption are used. They compare the behaviour of a particular surface to a theoretically perfect absorber and emitter called a 'black body', whose coefficient is given a value of one.

- **EMISSIVITY is the fraction of energy radiated by a body compared to that radiated by a black body at the same temperature**

Similarly, the absorptivity, or absorption factor, is the fraction of radiant energy absorbed by a body compared to that absorbed by a black body.

Values of emissivity and absorptivity depend upon the wavelength of the radiation and this is determined by the temperature of the source of the radiation. The Sun is a high temperature source of radiation and building materials are low temperature sources, so different sets of values may be quoted for the same surface. Table 2.2 gives typical values for common building surfaces.

In general, the colour of most building materials has an important effect on the heat absorbed by the building from the Sun (high temperature radiation) but has little effect on the heat emitted from buildings (low temperature radiation). Certain types of glazing are classified as 'low emissivity' when special coatings on the glass gives an emissivity coefficient of less than 0.2.

Table 2.2 *Surface coefficients for building materials*

Surface	Emissivity (low temperature radiation)	Absorptivity (solar radiation)
Aluminium	0.05	0.2
Asphalt	0.95	0.9
Brick – dark	0.9	0.6
Paint – white	0.9	0.3
Paint – black	0.9	0.9
Slate	0.9	0.0

Clear sky radiation

Clear sky radiation is a mechanism which can cause the structural temperature of a roof to fall significantly below the temperature of the surrounding air. At night time a building emits radiant heat to its surroundings and the rate of this heat loss from the roof will be increased if the night sky is clear and cloudless. This radiant 'suck' occurs because a clear dark sky is closer in form to a black body than is a cloudy sky and the clear sky therefore acts as a better absorber of radiant heat. The same radiant mechanism causes dew or ground frost to occur during a clear starry night.

U-VALUES

Thermal transmittance, *U*-value

Heat is transferred through an element of a building, such as a wall, by a number of mechanisms. Layers of different materials conduct heat at different rates; in any cavity there is heat transfer by conduction and convection in the air and by radiation effects. At the inside and outside surfaces of the wall the heat transfer by radiation and convection is affected by factors such as surface colour and exposure to climate.

It is convenient to combine all these factors in an ambitious single measurement which describes the air-to-air behaviour of a particular construction. This measurement is called the *overall thermal transmittance coefficient* or *U*-value.

- **A *U*-VALUE is a measure of the overall rate of heat transfer, by all mechanisms under standard conditions, through a particular section of construction**

  UNIT: $W/m^2 K$ ($W/m^2 °C$ may also be found in use)

The coefficient, or *U*-value, is measured as the rate of heat flow in watts through $1 m^2$ of a structure when there is a temperature difference across the structure of $1 °C$.

- Lower *U*-values provide better thermal insulation.

For example, a wall with a *U*-value of $0.4 W/m^2 K$ loses heat at half the rate of a wall with a *U*-value of $0.8 W/m^2 K$. Therefore, the cost of replacing heat lost through the first wall will be half the cost of that for the second wall.

Standard *U*-values

Standard *U*-values are calculated by making certain assumptions about moisture contents of materials and about rates of heat transfer at surfaces

Table 2.3 *Typical U-values of common constructions*

Element	Composition	U-value (W/m² K)
Solid wall	brickwork, 215 mm plaster, 15 mm	2.3
Cavity wall	brickwork, 102.5 mm unventilated cavity, 50 mm brickwork, 102.5 mm	1.6
Cavity wall	brickwork, 102.5 mm unventilated cavity, 50 mm lightweight concrete block, 100 mm lightweight plaster, 13 mm	0.96
Cavity wall	brickwork, 102.5 mm unventilated cavity, 25 mm polystyrene board, 25 mm lightweight concrete block, 100 mm lightweight plaster, 13 mm	0.58
Cavity wall	brickwork, 102.5 mm insulation-filled cavity*, 60 mm lightweight concrete block, 100 mm plasterboard, 13 mm	0.45
Timber frame wall	brickwork, 102.5 mm unventilated cavity insulating material*, 60 mm, airgap plasterboard, 13 mm	0.45
Pitched roof	tiles on battens and felt ventilated loft airspace mineral fibre, 150 mm plasterboard, 13 mm	0.25
Industrial roof	outer rain shield (negligible insulation) insulating material*, 65 mm internal roof space plasterboard or lining, 13 mm	0.45
Window	single glazing	5.7
Window	double glazing	2.8
Window	Window triple glazing or double glazing with low emissivity glass	2.0

Key
* 'Insulating materials' in this table are taken to have *k*-values of 0.04 W/m K or less.

Notes
Floors: *U*-values are affected by the length of perimeter exposed to the outside. See table 2.4.
Windows: these *U*-values are assumed unless manufacturer shows otherwise.
The calculation of *U*-values is treated later in the chapter.

and in cavities. Although the standard assumptions represent practical conditions as far as possible, they will not always agree exactly with U-values measured on site.

Standard U-values, however, are needed as a common basis for comparing the thermal insulation of different types of structure and for predicting the heat losses from buildings. Standard U-values are also used to specify the amount of thermal insulation required by clients or by regulations.

The U-values for some common types of construction are given in table 2.3.

Insulation regulations

A significant method of minimising the energy used in buildings is to enforce good standards of thermal insulation by means of target U-values. Figure 2.2 and table 2.4 give typical figures. The control of other factors should also be considered and is considered in the next chapter. These other factors include ventilation rates, internal air temperatures, efficiency of plant, and utilisation of buildings.

Table 2.4 *Typical insulation standards for dwellings*

Element	Standard U-values (W/m² K)	
	SAP 60 or less	over SAP 60
Roofs	0.2	0.25
Flat roof	0.35	0.35
Exposed walls	0.45	0.45
Ground floors/Exposed floors	0.35	0.35
Semi-exposed walls and floors	0.6	0.6
Windows, doors and rooflights	3.0	3.3
Maximum area double glazing	22.5% floor area	

Notes
The SAP (Standard Assessment Procedure) Energy Rating of a dwelling is based on annual energy costs (see chapter 3).
Areas of walls, roofs and floors are measured between internal faces.
Exposed elements are those exposed to outside air (including a suspended floor) or an element in contact with the ground.
Semi-exposed elements are those separating a heated space from an unheated space which does not meet standards.
The values are summarised from typical Building Regulations for England and Wales.
Such values are revised periodically and the latest regulations applicable to a particular case should be consulted.

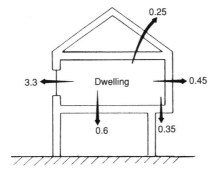

Figure 2.2 *Maximum U-values which apply to a building with an Energy Rating (SAP) greater than 60*

Thermal resistance, *R*-value

U-values are calculated from the thermal resistances of the parts making up a particular part of a structure. The different layers and surfaces of a building element, such as a wall, oppose the transmission of heat by varying amounts. These differences are described by thermal resistances, or *R-values*.

- ***THERMAL RESISTANCE (R)* is a measure of the opposition to heat transfer offered by a particular component in a building element**

  UNIT: $m^2 K/W$

  The idea of thermal resistance is comparable to electrical resistance. The term *Conductance (C)* is sometimes used to express the reciprocal of thermal resistance, where $C = 1/R$.

- Higher thermal resistance gives better thermal insulation.

There are three general types of thermal resistance which need to be determined, either by calculation or by seeking published standard values.

Material resistances
The thermal resistance of each layer of material in a structure depends on the rate at which the material conducts heat and the thickness of the material. Assuming that a material is homogeneous, this type of resistance can be calculated by the following formula.

$$R = d/k$$

where R = thermal resistance of that component ($m^2 K/W$)
      d = thickness of the material (m)
      k = thermal conductivity of the material (W/m K).

Alternatively

$$R = r \times d$$

where $r = 1/k$ = *resistivity* of that material (m K/W).

Surface resistances

The thermal resistance of an open surface depends upon the conduction, convection, and radiation at that surface. The air in contact with a surface forms a stationary layer which opposes the flow of heat. Surface resistances are usually found by consulting standard values which have been found by measurement or by advanced calculations. Some useful values are given in table 2.5.

Table 2.5 Standard thermal resistances

Type of resistance	Construction element	Heat flow	Surface emissivity*†	Standard resistances†† $(m^2 K/W)$
Inside surfaces	Walls	Horizontal	High	0.123
			Low	0.304
	Roofs pitched or flat Ceilings Floors	Upward	High	0.106
			Low	0.218
	Ceilings Floors	Downward	High	0.150
			Low	0.562
Outside surfaces (normal esposure)‡	Walls	Horizontal	High	0.055
			Low	0.067
	Roofs	Upward	High	0.045
			Low	0.053
Airspaces (including boundary surfaces)	Unventilated, 5 mm	Horizontal or Upward	High	0.11
			Low	0.18
	Unventilated, 20 mm or greater Ventilated loft space with flat ceiling, unsealed tiled pitched roof	Horizontal or Upward	High Low	0.18 0.35 0.11

Notes
* High emissivity is for all normal building materials, including regular glass.
† Low emissivity is for untreated metallic surfaces such as aluminium or galvanised steel, and for specially-coated glass.
‡ Normal exposure is for most suburban and country premises.
†† Further standard resistances are quoted in the *CIBSE Guide* or *BRE Digests*.

Factors which affect surface resistances are given below:

- Direction of heat flow: upward or downward
- Climatic effects: sheltered or exposed
- Surface properties: normal building materials with high emissivity or polished metal with low emissivity.

Airspace resistances

The thermal resistance of an airspace or empty cavity depends on the nature of any conduction, convection and radiation within the cavity. Airspace resistances are usually found by consulting published standard values. Some useful airspace resistances are given in table 2.5.

Factors which affect airspace resistances are given below:

- Thickness of the airspace
- Flow of air in airspace: ventilated or unventilated
- Lining of airspace: normal surfaces or reflective surfaces of low emissivity.

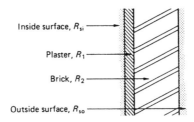

Figure 2.3 *Thermal resistances*

Total thermal resistance

The thermal resistances of the consecutive layers in a structural element, such as a wall or roof, can be likened to electrical resistances connected in series. Thus the total thermal resistance is the sum of the thermal resistances of all the components in a structural element.

Calculation of *U*-values

The thermal transmittance, or *U*-value, is calculated as the reciprocal of the total thermal resistance using the following formula:

$$U = \frac{1}{R_T}$$

UNIT: W/m^2 K

or

$$U = \frac{1}{R_{si} + R_1 + R_2 + \ldots + R_a + R_{so}}$$

where
U = U-value
R_T = sum of all component thermal resistances (R values)
R_{si} = standard inside surface resistance
R_1, R_2 = resistance of that particular material
R_a = standard resistance of any airspace
R_{so} = standard outside surface resistance.

Calculation guide
The following rules are useful in the calculation of practical U-values:

- Remember that there are always at least *two* surface resistances, the inside surface and the outside surface, even for the simplest element such as a window pane.
- Convert the thickness of all materials from millimetres to metres.
- Use the tables given later in the chapter to determine the U-values of ground floors. Only exposed floors, such as an overhang, can be calculated as in worked example 2.1.
- Work to a final accuracy of two decimal places. The k-values and standard resistances used in the calculations are not precise enough for higher accuracy.
- Ignore the effect of timber joists or frames, wall ties, thin cavity closures, damp-proof membranes and other thin components.
- Lay the calculation out in a table, as shown in worked example 2.1. This method encourages clear accurate working, allows easy checking, and is also suitable for transfer to a computer spreadsheet.
- List the layers of the structure in correct order from high temperature to low temperature. Although the order makes no difference to the simple addition of resistances, it will affect the later use of the calculation for temperature profiles.

Worked example 2.1
Calculate the U-value of a cavity wall with a 105 mm thick brick outer leaf, a 75 mm unventilated cavity containing 50 mm of fibreglass quilt, then a 100 mm lightweight concrete block inner leaf with a 15 mm layer of lightweight plaster. Thermal conductivities in W/m K are: brickwork 0.84, lightweight concrete blocks 0.19, lightweight plaster 0.16, fibreglass 0.04. Standard thermal resistances in m^2 K/W are: internal surface 0.123, external surface 0.055, cavity 0.18.

Step 1: Sketch a diagram indicating all parts of the construction with surface layers.

Step 2: Tabulate all information and, where necessary, calculate thermal resistance using $R = d/k$.

Layer	Thickness (m)	Thermal conductivity (W/m K)	Resistance (m² K/W)	
Internal surface	n/a	n/a	Standard	= 0.123
Lightweight plaster	0.015	0.16	0.015/0.16	= 0.094
Lightweight concrete block	0.100	0.19	0.1/0.19	= 0.526
Fibreglass quilt	0.050	0.04	0.05/0.04	= 1.25
Cavity	0.025	n/a	Standard	= 0.18
Exposed brickwork	0.105	0.84	0.105/0.84	= 0.125
External surface	n/a	n/a	Standard	= 0.055
			Total resistance, R_T	= 2.353

Step 3: Using $U = \dfrac{1}{R_T}$

$$U = \frac{1}{2.353} = 0.425$$

So U-value = **0.43 W/m² K**

U-values for floors

When heat leaves a building through a ground floor, the path of the heat flow has to curve through the ground back to the outside air at the edge of the building. Heat losses through the floor are therefore highest near the exposed edges of the floor, and heat losses are lowest near the centre of the floor where the ground itself adds to the insulation.

The calculation of the actual U-values for ground floors is complex and it is usual to use tables or graphs to find the minimum thickness of insulation needed to meet a particular average standard required by regulations. For some large buildings the standard may be achieved without the use of insulation because the ratio of exposed floor edges to total area is proportionally small.

In using tables for floors it is first necessary to calculate the floor perimeter to area ratio

P/A

where P = floor perimeter length (m)
A = area of floor (m^2).

Table 2.6 *Floor insulation*

SOLID FLOOR IN CONTACT WITH GROUND
Insulation Thickness (mm) to achieve U-value of
0.25 W/m^{2}K

| P/A | Thermal conductivity of insulant (W/mK) | | |
	0.02	0.035	0.05
1.00	62	108	155
0.60	56	98	139
0.20	24	42	60

Insulation Thickness (mm) to achieve U-value of
0.45 W/m^{2}K

| P/A | Thermal conductivity of insulant (W/mK) | | |
	0.02	0.035	0.05
1.00	26	46	66
0.60	20	35	50
0.30	4	6	9

SUSPENDED TIMBER FLOOR
Insulation Thickness (mm) to achieve U-value of
0.25 W/m^{2}K

| P/A | Thermal conductivity of insulant (W/mK) | | |
	0.02	0.035	0.05
1.00	95	140	184
0.60	85	126	166
0.20	33	52	69

Insulation Thickness (mm) to achieve U-value of
0.45 W/m^{2}K

| P/A | Thermal conductivity of insulant (W/mK) | | |
	0.02	0.035	0.05
1.00	37	57	77
0.60	27	43	58
0.30	4	7	10

Note: P/A is the ratio of the perimeter (m) to floor area (m^2).

Adjustments to *U*-values

It is sometimes necessary to calculate the effect that additional insulating material has upon a *U*-value, or to calculate the thickness of material that is required to produce a specified *U*-value. Use the following guidelines to make adjustments to *U*-values:

- *U*-values can *not* be added together or subtracted from one another
- Thermal resistances, however, can be added and subtracted. The resistances making up a particular *U*-value can then be adjusted to produce the new *U*-value
- Worked example 2.2 illustrates the technique.

Worked example 2.2

A certain uninsulated cavity wall has a *U*-value of 0.91 W/m² K. If expanded polyurethane board is included in the construction what minimum thickness of this board is needed to reduce the *U*-value to 0.45 W/m² K? Given that the thermal conductivity of the expanded polyurethane = 0.025 W/m K.

Target *U*-value	$U_2 = 0.45$
Target total resistance (1/*U*)	$R_2 = 1/0.45 = 2.222$
Existing *U*-value	$U_1 = 0.91$
Existing total resistance (1/*U*)	$R_1 = 1/0.91 = 1.099$
Extra resistance required	$R_2 - R_1 = 2.222 - 1.099$
	$= 1.123$

Know that the *k*-value of proposed insulating material $k = 0.025$, so using formula $\quad R = d/k$:

Thickness of material	$d = R \times k$
	$= 1.123 \times 0.025$
	$= 0.028$ metres

So minimum thickness of insulating board needed to give 0.45 *U*-value is **28 mm**

THERMAL BRIDGING

When a material of high thermal conductivity passes completely through a wall, floor, or roof then the insulation in that area is said to be 'bridged' and thus its effective *U*-value is reduced.

- **A *THERMAL BRIDGE* is a portion of a structure whose higher thermal conductivity lowers the overall thermal insulation of the structure**

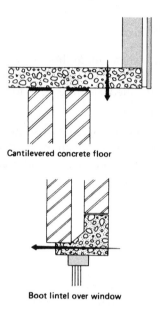

Cantilevered concrete floor

Boot lintel over window

Figure 2.4 *Thermal bridges*

There is increased heat flow across the thermal bridge and the surfaces on the interior side of the bridge therefore become cooler, giving rise to the informal term of 'cold bridge'. There is an increased risk of condensation and mould growth on these internal surfaces around cold bridges, such as the lintel above a window.

Thermal bridges can easily occur when the insulation of a wall is bridged at the junctions with the floor, the roof, or the windows. Some examples of troublesome thermal bridges are shown in figure 2.4. The remedy for thermal bridging is the correct design and installation of thermal insulation. Even in well-insulated modern construction some thermal bridging is inevitable when structural members, such as timber frames, must interrupt the main insulation material. The effects can be calculated by using the procedure shown in the section on *thermal bridge calculation.*

Average *U*-values

If a wall, or other element, is composed of different constructions with different *U*-values then the overall insulation of the wall depends upon the relative areas of the different constructions. For example, if a wall contains two-thirds brickwork and one-third windows then the *U*-values of the brickwork and windows must be combined in the proportions of 2/3 and 1/2 to give the average *U*-value for the wall.

The general formula is as follows:

$$U\left(\text{average}\right) = \frac{A_1U_1 + A_2U_2 + \dots}{A_1 + A_2 + \dots}$$

where A_1, A_2, etc. are the areas with the U-values U_1, U_2, etc.

Worked example 2.3

A wall has a total area of $8\,m^2$ of which $2\,m^2$ are windows. The U-values are $0.35\,W/m^2\,K$ for the cladding and $2.8\,W/m^2\,K$ for the glazing. Calculate the average U-value for the wall.

Know $U_1 = 0.35$, $A_1 = 8 - 2 = 6$
            $U_2 = 2.8$ $A_2 = 2$
U (average) $= ?$

Using $U = \dfrac{A_1U_1 + A_2U_2}{A_1 + A_2}$

$$U = \frac{\left(6 \times 0.35\right) + \left(2 \times 2.8\right)}{6 + 2} = \frac{2.1 + 5.6}{8} = 0.96$$

So average U-value = **$0.96\,W/m^2\,K$**

Thermal bridge calculation

To calculate the effect of thermal bridges a proportional method can also be used. Because the bridging might take place in one layer of the construction then a proportional method is applied to values of thermal resistance before a final U-value is calculated.

When the bridging is at regular intervals, such as with timber studs, joists, or mortar joint in lightweight block, it is common to use a standard table such as those in guides to the Building Regulations. These tables have been prepared by calculations, or by spreadsheets, based on examples such as worked example 2.4.

Worked example 2.4

An external wall has the following construction: outer leaf of brickwork; air cavity; inner leaf of ply sheathing, timber stud frame filled with insulation, plasterboard.

The inner leaf has a mixed construction with lower heat flow through the insulation areas and higher heat flow through the timber stud areas. Therefore the thermal resistance is calculated separately for each part of the inner leaf, as in the following.

Resistance through section containing timber stud m^2K/W
Inside surface resistance $= 0.12$
Resistance of plasterboard $= 0.013/0.16$ $= 0.08$
Resistance of timber stud $= 0.09/0.14$ $= 0.64$
Resistance of ply sheathing $= 0.009/0.14$ $= 0.06$
Half cavity resistance $= 1/2$ (std 0.18) $= 0.09$
Total resistance through timber stud section R_1 $= 0.99$

Resistance through section containing insulation
Inside surface resistance $= 0.12$
Resistance of plasterboard $= 0.013/0.16$ $= 0.08$
Resistance of insulation $= 0.09/0.04$ $= 2.25$
Resistance of ply sheathing $= 0.009/0.14$ $= 0.06$
Half cavity resistance $= 1/2$ (std 0.18) $= 0.09$
Total resistance through insulation section R_2 $= 2.60$

If the timber studs occur after every 600 mm of insulation and the studs are 38 mm wide, then the fractional areas are

for timber stud $F_1 = 38/600 = 0.063$
for insulation $F_2 = 1 - 0.063 = 0.937$

The resistance of the inner leaf is obtained from using the partial resistances and corresponding fractional areas in the following formula:

$$\frac{1}{R_3} = \frac{F_1}{R_1} + \frac{F_2}{R_2}$$

$$\frac{1}{R_3} = \frac{0.063}{0.99} + \frac{0.937}{2.6}$$

$$\frac{1}{R_3} = 0.064 + 0.36 = 0.424$$

$$\frac{1}{R_3} = \frac{1}{0.424} = 2.36$$

So resistance of the inner leaf is $2.36 \, m^2K/W$

This resistance for the inner leaf (and half the cavity) can then be added to the resistance for the outer leaf (including half the cavity) to give the total resistance for the wall. For example, if the resistance of the outer leaf has a typical value of $0.30 \, m^2K/W$, then

$$R_{total} = R_{inner} + R_{outer}$$
$$= 2.36 + 0.30$$
$$= 2.66 \, m^2K/W$$

so

$U = 1/R = 1/2.66 = 0.38$

So the U-value for the wall is **0.38 W/m²K**

Pattern staining

Pattern staining on a ceiling is the formation of a pattern, in dirt or dust, which outlines the hidden structure of the ceiling. It is a particular result of thermal bridging and also depends upon the frequency of redecoration. The areas of lower insulation transmit heat at a higher rate than the surrounding areas, such as between the joists shown in figure 2.5, for example. The convection patterns which then occur cause a greater accumulation of airborne dirt on these areas.

There is a risk of pattern staining occurring if the difference between surface temperatures exceeds 1 °C, which is unlikely in a modern well-insulated roof but can be seen in older buildings.

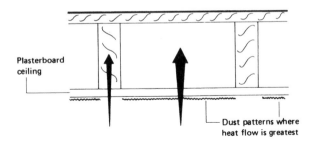

Figure 2.5 *Pattern staining*

STRUCTURAL TEMPERATURES

The thermal insulation installed in a building affects the rate at which the building loses heat energy which is measured by the U-value. The thermal performance of the building also depends upon the thermal capacity of the insulating material, which affects the times taken to heat or cool the structure, and the position of the insulation, which affects the temperatures in the structural elements.

Response times

It is possible for two types of wall to have the same thermal insulation, as measured by U-values, but to absorb or dissipate heat at different rates. As a result the rooms inside such walls would take different times to warm or to cool, as indicated in figure 2.6.

In general, lightweight structures respond more quickly to surrounding temperature changes than do heavyweight structures. This is because heavy-weight materials have a higher thermal capacity and require more heat energy to produce given temperature changes. These effects are described further in chapter 3.

The storage properties and slow temperature changes of heavyweight materials, such as concrete and brick, can be useful if thoughtfully designed. Where quick heating is often required, in bedrooms for example, a light-weight structure may be more useful. Or the heating-up time of heavyweight construction can be reduced by insulating the internal surfaces, such as by carpeting a concrete floor.

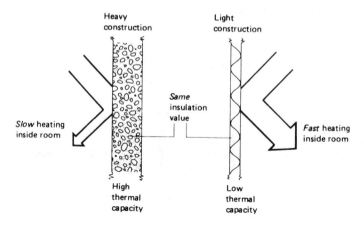

Figure 2.6 *Thermal response*

Temperature gradients

A temperature difference between the inside and outside of a wall or roof causes a progressive change in temperature from the warm side to the cold side. This *temperature gradient* changes uniformly through each compo-nent, provided that the material is homogeneous and that the temperatures remain constant.

A structure made up of different materials, such as the wall shown in figure 2.7, will have varying temperature gradients between the inside and

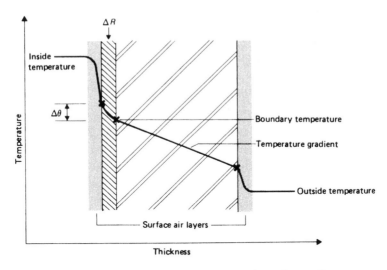

Figure 2.7 *Temperature gradients through a wall*

outside. The layers with the highest thermal resistances will have the steepest gradients. This is because the best insulators must have the greatest temperature differences between their surfaces.

The boundary temperatures between layers in a structural element can be determined from the thermal resistances which make up the *U*-value of that element.

The ratio of the temperature changes inside a structure is proportional to the ratio of the thermal resistances

The most useful relationship is given by the following formula:

$$\frac{\Delta\theta}{\theta_T} = \frac{R}{R_T}$$

where $\Delta\theta$ = temperature difference across a particular layer
R = resistance of that layer
θ_T = total temperature difference across the structure
R_T = total resistance of the structure.

From this relationship the temperatures at boundaries in a structure can be predicted for specified conditions. A calculation of all the boundary temperatures allows temperature gradients to be drawn on a scaled diagram. These diagrams can be used to predict zones of possible condensation, as shown in chapter 4.

Worked example 2.5

A room has an external wall with a U-value of 1.5 W/m K and contains air at 20 °C when the outside temperature is 5 °C. The internal surface resistance is 0.123 m² K/W. Calculate the boundary temperature on the internal surface of the wall.

Temperature drop across internal surface layer $\Delta\theta = ?$
Total temperature drop across wall $\qquad\qquad\theta_T = 20 - 5 = 15$
Resistance of internal surface layer $\qquad\qquad R_{si} = 0.123$
Total resistance of the wall $\qquad\qquad\qquad R_T = 1/U$
$\qquad\qquad\qquad\qquad\qquad\qquad\qquad\qquad\qquad = 1/1.5$
$\qquad\qquad\qquad\qquad\qquad\qquad\qquad\qquad\qquad = 0.667$

Using

$$\frac{\Delta\theta}{\theta_T} = \frac{R}{R_T}$$

$$\frac{\Delta\theta}{15} = \frac{0.123}{0.667}$$

$$\Delta\theta = \frac{0.123}{0.667} \times 15 = 2.77$$

Temperature drop across internal surface layer = 2.77 °C
So temperature on the inside surface = 20 − 2.77 = **17.23 °C**

Exercises

1 Choose three different insulating materials used in modern buildings. List the physical properties of each material and use the principles of heat transfer to explain why the material acts as a good thermal insulator. Use a construction sketch to help describe where and how the material is installed in a building.

2 Use definitions, units and examples to carefully explain the difference between the following terms:
   (a) thermal conductivity;
   (b) thermal resistance; and
   (c) thermal transmittance.

3 A cavity wall is constructed as follows: brickwork outer leaf 105 mm, air gap 25 mm, expanded polystyrene board 25 mm, lightweight concrete block inner leaf 100 mm, plasterboard 10 mm. The relevant values of thermal conductivity, in W/m K, are: brickwork 0.84, polystyrene 0.035, concrete block 0.19, plasterboard 0.16. The standard thermal

resistances, in $m^2 K/W$ are: outside surface 0.055, inside surface 0.123, air gap 0.18.

(a) Calculate the *U*-value of this wall.

(b) Calculate the *U*-value of the same wall sited in a position of 'severe exposure' for which the outside surface resistance is $0.03 \, m^2 K/W$.

4 The cavity wall of an existing house has outer and inner brickwork leaves each 105 mm with a 50 mm air gap between them, finished with a 16 mm layer of plaster inside. The relevant values of thermal conductivity, in W/mK are: brickwork 0.73, plaster 0.46. The standard thermal resistances, in $m^2 K/W$, are: outside surface 0.055, inside surface 0.123, air gap 0.18.

(a) Calculate the *U*-value of the existing wall.

(b) Calculate the *U*-value of the wall if the cavity is completely filled with foamed urea formaldehyde ($k = 0.026 \, W/mK$).

5 Compare the *U*-values obtained in questions 3 and 4 with the *U*-values required in current Building Regulations applicable to your area. Comment on the suitability of the walls for different purposes.

6 Compare the *U*-values of a single-glazed window made up of one sheet of 4 mm glass with a double glazed window made up of two sheets of 4 mm glass which have a 5 mm airspace between them. The thermal conductivity of the glass is 1.022 W/mK. The standard thermal resistances, in $m^2 K/W$, are: outside surface 0.055, inside surface 0.123, airspace 0.11. Comment on the proportion of the thermal resistance provided by the glass layers. Comment on the effect of the window frames.

7 A blockwork wall measures $5 \, m \times 2.8 \, m$ in overall length and height. The wall contains one window 1400 mm by 800 mm and one door 1900 mm by 750 mm. The *U*-values, in $W/m^2 K$, are: blockwork 0.58, window 5.6, door 3.4. Calculate the average *U*-value of this wall.

8 A wall panel is to have the following construction: outer metal sheeting, foamed polyurethane board, airgap, and 15 mm of lining board. The relevant values of thermal conductivity, in W/mK, are polyurethane board 0.025, and lining board 0.16 (the metal sheeting is ignored). The standard thermal resistances, in $m^2 K/W$ are: outside surface 0.055, inside surface 0.123, airgap 0.18. Calculate the minimum thickness of polystyrene needed to give the wall panel a *U*-value of $0.45 \, W/m^2 K$.

9 A domestic pitched roof of tiles on felt sacking, with a plasterboard ceiling, has an existing *U*-value of $1.9 \, W/m^2 K$. Calculate the minimum thickness of fibreglass insulation in the roof space required to give the roof a new *U*-value of $0.25 \, W/m^2 K$. The thermal conductivity of the fibreglass quilting used is 0.04 W/mK.

10 A wall has a U-value of 2.5 W/m²K. The thermal resistance of the inside surface layer is 0.123 m²K/W. The inside air temperature is 18°C and the outside air temperature is 0°C. Calculate the temperature on the inside surface of the wall.

Answers

3 (a) 0.56 W/m²K; **(b)** 0.57 W/m²K

4 (a) 1.47 W/m²K; **(b)** 0.41 W/m²K

6 5.49 W/m²K; 3.38 W/m²K

7 1.27 W/m²K

8 44 mm

9 139 mm

10 12.5°C

3 Energy Use

A satisfactory thermal environment is an important purpose of good building design. To achieve an acceptable thermal environment we need to consider the comfort of people using the building and the requirements of objects stored in the building.

The mechanisms causing heat to be lost from the building can be identified and the resulting heat losses then calculated. Heat gains, such as those from the Sun or from electrical appliances, also need to be taken into account. From such considerations it is possible to design the correct type of heating and cooling plant for the building, to predict the energy consumption and to calculate the running costs.

THERMAL COMFORT

The thermal comfort of human beings is governed by many physiological mechanisms of the body and these vary from person to person. In any particular thermal environment it is difficult to get more than 50 per cent of the people affected to agree that the conditions are comfortable!

The body constantly produces heat energy from the food energy it consumes. This heat needs to be dissipated at an appropriate rate to keep the body at constant temperature. The transfer of the heat from the body is mainly by the processes of convection, radiation, and evaporation. Evaporation transfers the latent heat we give to the water vapour which is given out on the skin (perspiration) and in the breath (respiration).

The total quantity of heat produced by a person depends upon the size, the age, the sex, the activity, and the clothing of the person. A further complication is the ability of the body to become accustomed to the surrounding conditions and to adapt to them. For example, everyone can tolerate slightly lower temperatures during winter. This adaptation can be influenced by the type of climate and the social habits of a country.

Factors affecting thermal comfort

The principal factors affecting thermal comfort can be conveniently considered under the headings below and are discussed further in the following sections.

Personal variables
- Activity
- Clothing
- Age
- Sex.

Physical variables
- Air temperature
- Surface temperatures
- Air movement
- Humidity.

Activity

The greater the activity of the body the more heat it gives off. The rate of heat emission depends upon the individual metabolic rate of a person and upon their surface area. People who seem similar in all other respects can vary by 10 to 20 per cent in their heat output.

The average rate of heat emission decreases with age. Table 3.1 lists typical heat outputs from an adult male for a number of different activities. The output from adult females is about 85 per cent that of males.

Table 3.1 *Typical heat output of human body*

Activity	Example	Typical heat emission of adult male
Immobile	Sleeping	70 W
Seated	Watching television	115 W
Light work	Office	140 W
Medium work	Factory, dancing	265 W
Heavy work	Lifting	440 W

Note: Adapted from the *CIBSE Guide.*

Clothing

Clothes act as a thermal insulator for the body and help to maintain the skin at a comfortable temperature. Variations in clothing have a significant effect on the surrounding temperatures that are required for comfort.

To enable heating needs to be predicted a scale of clothing has been developed, the clo-value. 1 clo represents $0.155\,m^2\,K/W$ of insulation and values range from 0 clo to 4 clo. Table 3.2 shows the value of different types of clothing and indicates how the room temperature required for comfort varies with clothing. On average, women prefer slightly higher temperatures.

Table 3.2 *Clothing values*

clo value	Clothing example	Typical comfort temperature when sitting (0 °C)
0 clo	Naked; swimwear	29
0.5 clo	Light trousers, shirt; light dress, blouse	25
1.0 clo	Business suit; dress, jumper	22
2.0 clo	Heavy suit, overcoat, gloves, hat	14

Note: Appropriate underclothing is assumed in each case.

Room temperatures

The temperature of the surrounding surfaces can affect the thermal comfort of people as much as the temperature of the surrounding air. This is because the rate at which heat is radiated from a person is affected by the radiant properties of the surroundings. For example, when sitting near the cold surface of a window the heat radiated from the body increases and can cause discomfort.

A satisfactory design temperature for achieving thermal comfort needs to take account of both air temperatures and radiant effects. Different types of temperatures are described below.

Inside air temperature t_{ai}
The inside air temperature is the average temperature of the bulk air inside a room. It is usually measured by an ordinary dry bulb thermometer which is suspended in the centre of the space and shielded from radiation.

Mean radiant temperature t_r
The mean radiant temperature is the average effect of radiation from surrounding surfaces. At the centre of the room this temperature can be taken as being equal to the *mean surface temperature* as calculated by

$$t_r = \frac{A_1 t_1 + A_2 t_2 + \ldots}{A_1 + A_2 + \ldots}$$

where $t_1, t_2 \ldots$ are the surface temperatures of the areas $A_1, A_2 \ldots$, etc.

The mean radiant temperature should be kept near the air temperature but not more than about 3 °C below it, otherwise conditions are sensed as 'stuffy'.

Inside environmental temperature t_{ei}

The environmental temperature is a combination of air temperature and radiant temperature. The exact value depends upon convection and radiation effects. For average conditions t_{ei} can be derived from the following formula:

$$t_{ei} = 2/3 \ t_r + 1/3 \ t_{ai}$$

Environmental temperature is recommended for the calculation of heat losses and energy requirements.

Dry resultant temperature t_{res}

Dry resultant temperature is a combination of air temperature, radiant temperature and air movement. When the air movement is low t_{res} can be derived from the following formula:

$$t_{res} = 1/2 \ t_r + 1/2 \ t_{ai}$$

Room centre comfort temperature t_c

A comfort temperature is a measure of temperature which gives an acceptable agreement with thermal comfort. When air movement is low, the dry resultant temperature at the centre of a room is a commonly-used comfort temperature.

The *globe thermometer* is a regular thermometer fixed inside a blackened globe of specified diameter (150 mm is one standard). This globe temperature can be used to calculate other temperatures and when air movement is small it approximates to the comfort temperature.

Table 3.3 *Typical design temperatures and infiltration rates*

Type of building	Design temperature t_{ei}	Infiltration, air changes per hour*
Domestic		
living rooms	21	1
bedrooms	18	1/2
bathrooms	22	1/2
Offices, general	20	1
Classrooms, school	18	2
Shops, large	18	1/2
Restaurants; bars	18	1
Hotel bedrooms	22	1
Factories, light work	16	

* Conversion from air changes per hour to ventilation heat loss is: 1 air change per hour = $0.34 \ W/m^2 \ °C$.

Air movement

The movement of air in a room helps to increase heat lost from the body by convection and can cause the sensation of draughts. The back of the neck, the forehead and the ankles are the most sensitive areas for chilling.

- Air movements above 0.1 m/s in speed require higher air temperatures to give the same degree of comfort

For example, if air at 18 °C increases in movement from 0.1 m/s to 0.2 m/s then the temperature of this air needs to rise to 21 °C to avoid discomfort.

The air movement rate is not the same thing as the air change rate and is not always caused by ventilation. Uncomfortable air movement may be due to natural convection currents, especially near windows or in rooms with high ceilings.

A *hot-wire anemometer* or a *Kata thermometer* may be used to measure air movement. Both devices make use of the cooling effect of moving air upon a thermometer.

Humidity

Humidity is caused by moisture in the air and will be treated more fully in later chapters.

- Relative humidity within the range of 40 to 70 per cent is required for comfortable conditions.

High humidities and high temperatures feel oppressive and natural cooling by perspiration is decreased. High humidity and low temperatures cause the air to feel chilly.

Low humidities can cause dryness of throats and skin. Static electricity can accumulate with low humidity, especially in modern offices with synthetic carpet, and cause mild but uncomfortable electric shocks.

Ventilation

In any occupied space ventilation is necessary to provide oxygen and to remove contaminated air. Fresh air contains about 21 per cent oxygen and 0.04 per cent carbon dioxide while expired air contains about 16 per cent oxygen and 4 per cent carbon dioxide. The body requires a constant supply of oxygen but the air would be unacceptable well before there was a danger to life. As well as being a comfort consideration the rate of ventilation has a great effect on the heat loss from buildings and on condensation in buildings.

Table 3.4 *Typical fresh air-supply rates*

Type of space	Recommended air-supply
Residences, offices, shops	8 litres/s per person
Restaurants, bars	18 litres/s per person
Kitchens, domestic toilets	10 litres/s per m² floor

The normal process of breathing gives significant quantities of latent heat and water vapour to the air. Household air is also contaminated by body odours, bacteria, and the products of smoking, cooking and washing. In places of work, contamination may be increased by a variety of gases and dusts.

A number of statutory regulations specify minimum rates of air-supply in occupied spaces. Recommended rates of ventilation depend upon the volume of a room, the number of occupants, the type of activity, and whether smoking is expected. It is difficult therefore to summarise figures for air-supply but table 3.4 quotes some typical values.

HEAT LOSSES

Factors affecting heat loss

Heat loss from a building occurs by a number of mechanisms, as illustrated in figure 3.1. Some important factors which affect the rate at which this heat is lost are listed here and summarised below:

- Insulation of building
- Area of the external shell
- Temperature difference
- Air change rate
- Exposure to climate
- Efficiency of services
- Use of building.

Insulation of shell
The heat loss from a building decreases as the insulation of the external fabric of the building is increased. The external parts of the structure surrounding occupied areas need most consideration but all buildings which are heated, for whatever purpose, should be well-insulated in order to save energy. The thermal transmittance coefficient, the *U*-value, of a construction is a commonly-used measure of insulation.

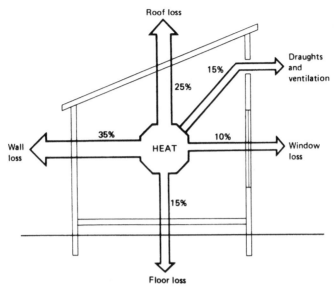

Approximate percentages for an uninsulated house

Figure 3.1 *Heat losses in a building*

Area of the shell

The greater the area of external surfaces the greater is the rate of heat loss from the building. A terraced house, for example, loses less heat than a detached house of similar size. Table 3.5 compares exposed perimeter areas for different shapes of dwelling.

The basic plan shape of a building is one of the first design decisions to be made, although choices may be restricted by the nature of the site and by local regulations.

Table 3.5 *Exposed areas of dwellings*

Type of dwelling (each of same floor area)	Exposed perimeter area (per cent)
Detached house	100
Semi-detached house	81
Terraced house	63
Flat on middle storey (two external walls)	32

Temperature difference

A large difference between the temperatures inside and outside the building increases the rate of heat lost by conduction and ventilation. This loss is affected by the design temperature of the inside air, which depends upon the purpose of the building. Recommended comfort temperatures for different types of buildings are given in table 3.3.

Air change rate

Warm air leaving a building carries heat and is replaced by colder air. The air flow occurs through windows, doors, gaps in construction, ventilators, and flues. This air change may be controlled ventilation or it may be accidental infiltration.

The rate of air change is also affected by effects of wind upon the building. Table 3.3 gives typical rates of air infiltration which are found to exist in buildings on sites of normal exposure during winter heating conditions.

Exposure to climate

When a wind blows across a wall or roof surface the rate of heat transfer through that element increases. This effect is included in the standard value of external surface resistance used in calculating a *U*-value. Standard surface resistances are available for three types of exposure listed below:

- **Sheltered:** Buildings up to three storeys in city centres
- **Normal:** Most suburban and country buildings
- **Severe:** Buildings on exposed hills or coastal sites. Floors above the fifth in suburban or country sites. Floors above the ninth in city centres.

Efficiency of services

There is usually some wastage of heat energy used for water heating and space heating, and the design of the services can minimise or make use of this waste heat. For example, some of the heat from the hot gases passing up a flue can help heat the building if the flue is positioned inside the building rather than on an external wall. More advanced techniques include the recovery of the latent heat contained in the moist flue gases produced by gas burners.

The heat given off by hot water storage cylinders and distribution pipes, even well-insulated ones, should be used inside the building if possible rather than wasted outside. Even the draining away of hot washing water, which is dirty but full of heat energy, is a heat loss from the building and a heat gain to the sewerage system! Methods of heat recovery are discussed in later sections on energy conservation.

Patterns of use

The number of hours per day and the days per year that a building is used have a large effect on the energy consumption of a building. Many buildings which are unoccupied at times, such as nights, need to be pre-heated before occupancy each morning.

These patterns of building use and occupancy vary greatly, even for similar buildings. When a building has separate areas with different patterns of occupancy, each part needs to be considered as a separate building for heating calculations.

Calculation of heat loss

Various methods are available for calculating the rate at which heat flows out of a building, and the quantity of heat lost in a given time. It is relatively difficult to calculate heat losses for unsteady or *cyclic* conditions where temperatures fluctuate with time. However, certain simplified calculations can be used for predicting heating requirements and the amount of fuel required. The results obtained by these calculations are found to give adequate agreement with the conditions that actually exist.

With *steady state* conditions the temperatures inside and outside the building do not change with time and the various flows of heat from the building occur at constant rates. Assuming steady state conditions the heat losses from a building can be classed as either a 'fabric loss' or a 'ventilation loss' and then calculated by the methods described below.

Conditions not suited to the steady state assumption occur when there is intermittent heating, such as that given by electric storage heaters or by solar radiation. The thermal capacity of the structure is then significant and needs to be considered in the calculation, as explained in the section about *admittance.*

Fabric heat loss

Fabric heat loss from a building is caused by the transmission of heat through the materials of walls, roofs and floors. Assuming steady state conditions, the heat loss for each element can be calculated by the following formula:

$$P_f = U \, A \, \Delta t$$

where P_f = rate of fabric heat loss
  = heat energy lost/time (W)
 U = U-value of the element considered (W/m^{2}K)
 A = area of that element (m^2)
 Δt = difference between the temperatures assumed for the inside and outside environments (°C).

The heat loss per second is a form of power (energy divided by time) and therefore measured in watts (which are joules per second). The notation P is used here to represent this rate of heat energy. Some CIBSE documents use the less-correct notation Q for rate of heat loss.

To calculate daily heat losses, appropriate temperatures would be the internal environmental temperature and the outside environmental temperature, both averaged over 24 hours. For the calculation of maximum heat losses, such as when choosing the size of heating equipment, it is necessary to assume a lowest external design temperature, such as $-1\,°C$.

Ventilation loss

Ventilation heat loss from a building is caused by the loss of warm air and its replacement by air that is colder and has to be heated. The rate of heat loss by such ventilation or infiltration is given by the following formula:

$$P_v = \frac{c_v N V \Delta t}{3600}$$

where P_v = rate of ventilation heat loss
$\qquad$ = heat energy/time (W)
$\quad c_v$ = volumetric specific heat capacity of air
$\qquad$ = specific heat capacity $\times$ density $(J/m^3 K)$
$\quad N$ = air infiltration rate for the room (the number of complete air changes per hour)
$\quad V$ = volume of the room (m^3)
$\quad \Delta t$ = difference between the inside and outside air temperatures $(°C)$.

The heat loss per second is a form of power (energy divided by time) and therefore measured in watts (which are joules per second). The notation P is used here to represent this rate of heat energy. Some CIBSE documents use the less-correct notation Q for rate of heat loss.

The values for the specific heat capacity and seconds in an hour are sometimes combined into a factor of 0.33, to give the following alternative formula:

$$P_v = 0.33\ N\ V\ \Delta t$$

To calculate daily heat losses appropriate temperatures would be the mean internal air temperature and the mean external air temperature, both averaged over 24 hours. For the calculation of maximum heat losses, such as when choosing the size of heating equipment, it is necessary to assume a lowest design temperature for the external air, such as $-1\,°C$.

External temperature

When designing heating systems for buildings it is necessary to assume a temperature for the outside environment. For winter heating an overcast sky is assumed and the outside air temperature can be used for design purposes. For heat transfer calculations in summer it is necessary to take account of solar radiation as well as air temperature.

Sol-air temperature t_{eo}

The sol-air temperature is an environmental temperature for the outside air which includes the effect of solar radiation:

- The rate of heat flow due to the *sol-air temperature* is equivalent to the rate of heat flow due to the actual air temperature combined with the effect of solar radiation.

Sol-air temperature varies with climate, time of day and incident radiation. Values can be calculated or found from tables of average values.

Worked example 3.1

A window measuring 2 m by 1.25 m has an average *U*-value, including the frame, of 6.2 W/m²K. Calculate the rate of fabric heat loss through this window when the inside comfort temperature is 20°C and the outside air temperature is 4°C.

Know

$$U = 6.2 \, \text{W/m}^2\text{K}, \, A = 2 \times 1.25 = 2.5 \, \text{m}^2, \, \Delta t = 20 - 4 = 16°\text{C}.$$

Using

$$P_i = U \, A \, \Delta t$$
$$= 6.2 \times 2.5 \times 16 = 248$$

So fabric loss = **248 W**

Worked example 3.2

A simple building is 4 m long by 3 m wide by 2.5 m high. In the walls there are two windows, each 1 m by 0.6 m, and there is one large door 1.75 m by 0.8 m. The construction has the following *U*-values in W/m²K: windows 5.6, door 2.0, walls 2.5, roof 3.0, floor 1.5. The inside environmental or comfort temperature is maintained at 18°C while the outside air temperature is 6°C. The volumetric specific heat capacity of the air is taken to be 1300 J/m°C. There are 1.5 air changes per hour. Calculate the total rate of heat loss for the building under the above conditions.

Step 1: Sketch the building, with its dimensions, as in figure 3.2. Calculate the areas and temperature differences.

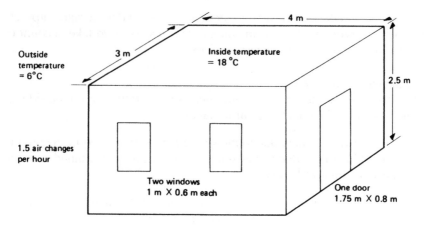

Figure 3.2 *Worked example 3.2*

Step 2: Tabulate the information and calculate the rate of fabric heat losses using $P_v = U A \Delta t$:

Element	U-value (W/m²K)	Area difference (m²)	Temperature (°C)	Rate of heat loss (W)
Window	5.6	1.2	12	80.64
Door	2.0	1.4	12	33.6
Walls	2.5	35–2.6	12	972
Roof	3.0	12	12	432
Floor	1.5	12	12	216
			Total rate of fabric heat loss	1734.24 W

Step 3: Calculate the ventilation heat loss.

$C_v = 1300 \, \text{J/m}^3\text{K}, \; N = 1.5 \, \text{h}^{-1}$
$V = 4 \times 3 \times 2.5 = 30 \, \text{m}^3, \; \Delta t = 18 - 6 = 12 \, °\text{C}$

Using

$$P_v = \frac{C_v \, N \, V \Delta t}{3600}$$

$$= \frac{1300 \times 1.5 \times 30 \times 12}{3600}$$

So rate of ventilation heat loss = 195 W

Step 4:
Total rate of heat loss = fabric heat loss + ventilation heat loss
$$= 1734.24 + 195$$
$$= \mathbf{1929.24\,W}$$

Non-steady conditions

For situations where the steady state assumption is invalid, it is necessary to consider the effects of cyclic (daily) variations in the outside temperature, variations in solar radiation, and changes in the internal heat input.

Thermal admittance
The thermal admittance value is one method of describing variations in temperature:

- ***Thermal admittance* or *Y-value* is a property of an element or a room which controls fluctuations in the inside temperature**

  UNIT: $W/m^2\,K$

The greater the admittance, the smaller the temperature swing inside the building. The factors which affect the admittance values of a particular

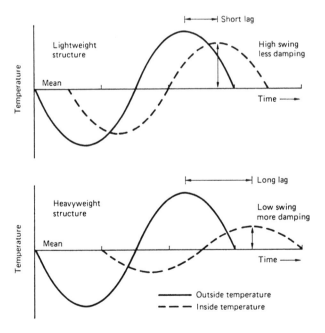

Figure 3.3 *Thermal response*

Table 3.6 *Contrasting values of admittance and transmittance*

Element	Y-value	U-value
Typical heavyweight wall (brick/blockwork with cavity insulation)	4.0	0.6
Typical lightweight wall (cladding, insulation, lining)	1.0	0.6

material include thermal conductivity and specific heat capacity. Methods of calculation using admittance values are available.

Dense heavyweight materials, such as concrete, have larger admittance values than the values for less dense materials, such as lightweight thermal insulating materials. Therefore, heavyweight structures have smaller temperature swings than lightweight structures.

Although the unit for admittance value is the same as for U-value, it is possible to have elements of different construction with identical insulation measured in U-values but different thermal damping properties as measured in admittance values. For very thin units, such as glass, the admittance becomes the same as the U-value.

HEAT GAINS

A building gains heat energy as well as loses heat energy, and both processes usually occur at the same time. In a location with a temperate climate such as in the British Isles, the overall gains are less than the overall losses, but the heat gains may still provide useful energy savings. The factors affecting heat gains are indicated in figure 3.4 and considered under the following categories:

- Solar heat gains from the Sun
- Casual heat gains from occupants and equipment in the building.

Solar heat gain

The heat gained in a building by radiation from the Sun depends upon the following factors:

- The geographical latitude of the site, which determines the height of the Sun in the sky
- The orientation of the building on the site, such as whether rooms are facing south or north
- The season of the year, which also affects the height of the Sun in the sky
- The local cloud conditions, which can block solar radiation

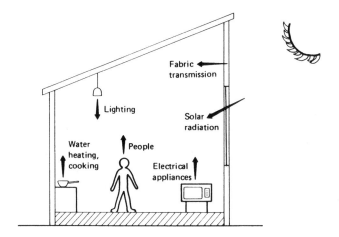

Figure 3.4 *Typical heat gains in a building*

- The angles between the Sun and the building surfaces, because maximum gain occurs when surfaces are at right angles to the rays from the Sun
- The nature of the window glass and whether it absorbs or reflects any radiation
- The nature of the roof and walls, because heavyweight materials behave differently to lightweight materials.

The rate at which heat from the Sun falls on a surface varies throughout the day and the year. Figure 3.5 gives an indication of the intensity of solar radiation on a vertical and horizontal surface at different times. The figures are for London (latitude 51.7 °N) and assume cloudless days. The heat gains for other surfaces and locations can be calculated from appropriate tables and charts. Further data about solar radiation is given in the section on climate in chapter 13.

Most solar heat gain to buildings in the United Kingdom is by direct radiation through windows. The maximum gains through south-facing windows tend to occur in spring and autumn when the lower angle of the Sun causes radiation to fall more directly onto vertical surfaces. This heat gain via windows can, if used correctly, be useful for winter heating.

The fabric solar heat gains through walls and roofs are considered negligible for masonry buildings during the United Kingdom winter. Little solar heat reaches the interior of the building because the high thermal capacity of heavyweight construction tends to delay transmission of the heat until its direction of flow is reversed with the arrival of evening.

The solar heat gains for a particular building at a specific time are relatively complicated to calculate, although it is important to do so when predicting summer heat gains in commercial buildings. For winter calcula-

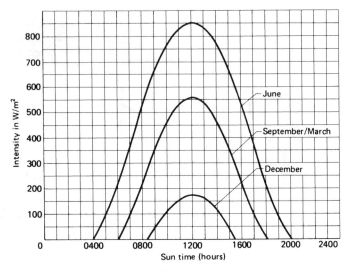

(a) Horizontal Surface

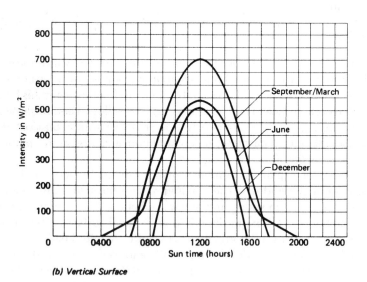

(b) Vertical Surface

Figure 3.5 *Solar intensities, latitute 51.7°N*

Table 3.7 *Seasonal solar gain through windows*

Type of window (unobstructed)	Seasonal total heat gain
South-facing windows	680 MJ/m² glass
East and west windows	410 MJ/m² glass
North-facing windows	250 MJ/m² glass
Total for average semi-detached house	7500 MJ

Note: seasonal figures are for 33 weeks.

tions however, it is useful to consider the total solar gain over an average heating season. Typical figures for solar heat gain through the windows of typical buildings in the UK are given in table 3.7.

Sun controls

Sun controls are parts of a building that help prevent excessive heat gain and glare caused by direct sunshine. The main types of device are described below:

- *External controls* are the most effective form of Sun control because they minimise the radiant heat reaching the fabric of the building. Examples include external shutters, awnings, projecting eaves or floor slabs.
- *Internal controls* such as blinds give protection against glare and direct radiation. The system is less effective than controls outside the glass because the blind will absorb some solar heat and re-emit this heat into the room. Examples include curtains, blinds, and internal shutters.
- *Special glasses* are available which prevent the transmission of most heat radiation with only some loss of light transmission. A similar effect is given by special film which sticks onto plain glass.

Casual heat gains

Casual heat gains take account of the heat given off by various activities and equipment in a building that are not primarily designed to give heat. The major sources of such heat are as follows:

- Heat from people
- Heat from lighting
- Heat from cooking and water heating
- Heat from machinery, refrigerators, electrical appliances.

In commercial or public buildings this type of heat gain must be allowed for in the design of the heating/cooling system. Where possible this heat

Table 3.8 *Heat emissions from casual sources*

Type of source	Typical heat emission
Adult person (for 20 °C surroundings)	
Seated at rest	90 W
Walking slowly	110 W
Medium work	140 W
Heavy work	190 W
Lighting	
Fluorescent system giving 400 lux (e.g. classroom)	20 W/m² floor area
Tungsten system giving 200 lux (e.g. domestic kitchen)	40 W/m² floor area
Equipment	
Desktop computer	150 W
Computer printer	100 W
Photocopier	800 W
Small Bunsen burner	600 W
Hair dryer	800 W
Gas cooker	3500 W per burner
Domestic fridge-freezer	150 W
Colour television, or hi-fi unit	100 W

Table 3.9 *Domestic seasonal heat gains*

Source	Typical gain per heating season
Body heat (per person)	1000 MJ
Cooking (gas)	6500 MJ
Cooking (electric)	3500 MJ
Water heating	2000 MJ
Electricity (including lights)	3000 MJ

Note: Figures for semi-detached house with 100 m². Seasonal results are for 33 weeks.

should be used rather than wasted. Table 3.8 gives typical rates of heat output from various sources.

In houses the casual heat gain is useful in winter and, as dwellings become better insulated, it forms a higher proportion of the total heat needed. The combined heat output of the various sources varies from hour to hour so it is again useful to consider the total heat gain over a standard heating system.

Table 3.9 quotes estimates of the heat gains in a typical semi-detached house, with a floor area of 100 m², totalled over a heating season of 33 weeks.

HEAT BALANCE

The thermal comfort of humans requires that the inside temperature of a building is kept constant at a specified level, and the storage of goods also needs constant temperatures. In order to maintain constant temperature the building will generally require heating or cooling, and both of these processes involve the consumption of energy.

Calculation of energy

Heat is defined as a form of energy. Power is defined as the rate (divide by time) at which energy used. The rate at which heat energy is used, measured in watts, may therefore be called a power loss. These power losses in watts are sometimes, and less accurately, called 'heat losses'.

The true heat or energy use can only be determined when it is decided which period of time is being considered. The quantity of energy used over a given period depends upon both the power (rate of using energy) and upon the time involved.

By definition

Energy = Power × Time

Written as a formula:

$$E = Pt$$

UNIT: joule (J), where 1 joule = 1 watt × 1 second.

There are some older, non-standard, units of energy which may be found in use:

kilowatt hour (kWh), where   1 kWh = 3.6 MJ
British Thermal Unit (BTU), where 1 BTU = 1.055 kJ

Energy balance

When heat losses and heat gains have been determined it is possible to calculate the extra energy needed to 'balance' the losses and the gains and to give a constant temperature. The following is a general expression of balance which is true for summer and winter conditions:

Fabric heat losses	Ventilation + heat losses	Solar = heat gains	Casual + heat gains	Energy for + heating or cooling

The quantities can be measured as true heat energy in joules or as rate of energy use in watts, but not by a mixture of methods. For most buildings in winter the 'useful energy' which needs to be supplied by the heating plant is given by the following expression:

Energy needed = Heat losses − Heat gains

Seasonal energy requirements

The energy requirement of a building at any particular time depends on the state of the heat losses and the heat gains at that time. These factors vary but, as described in the previous sections, it is useful to consider the total effect over a standard heating season.

It is important to note that the calculation of seasonal heat losses and gains assumes average temperature conditions and cannot be used to predict the size of the heating or cooling plant required; such a prediction needs consideration of the coldest and hottest days. Seasonal heat calculations are, however, valid for calculating total energy consumption and can therefore be used to predict the quantity of fuel required in a season and how much it will cost.

Worked example 3.3

Over a heating season of 33 weeks the average rate of heat loss from a certain semi-detached house is 2500 W for the fabric loss and 1300 W for the ventilation loss. The windows have areas: $6\,m^2$ south-facing, $5\,m^2$ east-facing, $6\,m^2$ north-facing. The house is occupied by three people and cooking is by gas.

Use the values for seasonal heat gains given in tables 3.7 and 3.9 and calculate:

(a) the seasonal heat losses;
(b) the seasonal heat gains; and
(c) the seasonal heat requirements.

(a) Total rate of heat loss = fabric loss + ventilation loss
$$= 2500\,W + 1300\,W$$
$$= 3800\,W$$

Heat energy lost = rate of heat loss × time taken
$$= 3800\,W \times (33 \times 24 \times 7 \times 60 \times 60)\,s$$
$$= 7.5842 \times 10^{10}$$
$$= 75.842\,GJ$$

So seasonal heat loss = **75 842 MJ**

(b) Heat gains

Solar window gain (table 3.7)	MJ
south ($680\,MJ/m^2 \times 6$)	4080
east ($410\,MJ/m^2 \times 5$)	2050
north ($250\,MJ/m^2 \times 6$)	1500

Casual gains (table 3.9)   MJ
   body heat (1000 MJ × 3)   3000
   cooking (gas)   6500
   water heating   2000
   electrical   3000
                          22 130 MJ

So seasonal heat gain = **22 130 MJ**

(c) Seasonal heat requirements = Heat loss − Heat gain
                                = 75 842 − 22 130 = 53 712 MJ
                                = **53.712 GJ**

ENERGY CONSUMPTION

Buildings in the British Isles, and in countries with a similar climate, suffer an overall loss of heat during the year. The energy required to replace heat losses from buildings represents a major portion of the country's total energy consumption. Figure 3.6 gives an indication of energy use in the United Kingdom in recent years. Building services make up 40 to 50 per cent of the national consumption of primary energy and about half of this energy is used in domestic buildings.

This section deals with the terminology and calculation of the energy used in buildings. The conservation of this energy used in buildings is also discussed in chapter 13.

Fuels

Almost all energy in common use comes from the Sun, even if it has been stored by various processes. Plants and trees use solar energy to grow material which can be burned directly, or decomposed into gases, or preserved underground. The energy in waves, wind and ocean heat has also been provided by the Sun.

- **A *Fuel* is a substance which is a source of energy**

Fossil fuels such as coal, crude oil and natural gas were formed by the decomposition of prehistoric organisms such as plants during geological changes in the Earth. Fossil fuels are *non-renewable energy sources* because the decomposition process has taken millions of years. Wind power, wave power, and similar processes are considered to be *renewable energy sources*.

Nuclear energy is released from small changes in the mass of sub-atomic particles during the processes of fission or fusion. Uranium and other mate-

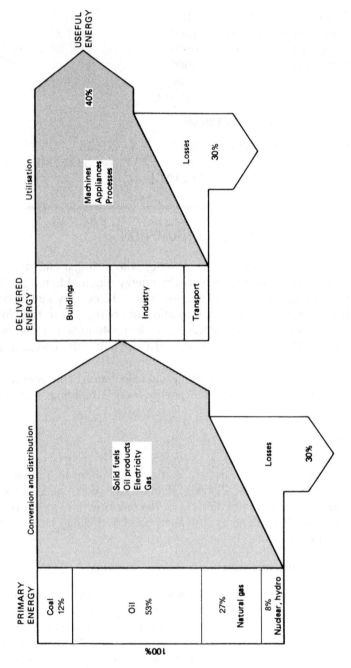

Figure 3.6 *Typical energy utilisation in the United Kingdom*

rials can be used as a fuel for generating such nuclear reactions. This process is also responsible for heat being generated within the Earth. The Sun generates its own heat by a fission reaction which depends on supplies of hydrogen which will eventually be depleted.

Energy terms

It is not possible to extract all the energy contained in a fuel and extra terms are used to describe the amounts of energy being supplied or paid for.

Primary energy
- *Primary energy* **is the total energy contained in fossil fuels such as coal, oil or natural gas**

The primary energy of a fuel is measured in the 'raw' state, before any energy is used in conversion or distribution.

Secondary energy
- *Secondary energy* **is the energy contained in a fuel which results from a conversion process**

Examples of secondary energy sources include electricity, manufactured gas, and surplus hot water. The secondary energy of electricity is only around 30 per cent of the primary energy contained in the original fuels; the rest is lost in the conversion and distribution processes.

Delivered energy
- *Delivered energy* **is energy content as it is received by the consumer**

It is the delivered energy or 'site energy' for which the consumer generally pays money, although some of this energy will be lost by conversion processes within the building.

Useful energy
- *Useful energy* **is the energy required to perform a given task**

It is useful energy that is needed to balance the heat losses and heat gains in a typical calculation.

Efficiency

The heat energy required for buildings is commonly obtained from fuels such as coal, gas and oil, even if the energy is delivered in the form of electricity. Each type of fuel must be converted to heat in an appropriate piece of equipment and the heat distributed as required.

The amount of heat finally obtained depends upon the original heat content of the fuel and the efficiency of the system in converting and distributing this heat.

Calorific value

- **Calorific value** **is a measure of the primary heat energy content of a fuel expressed in terms of unit mass or volume**

Some typical calorific values are quoted in table 3.10. These values may be used to predict the quantity of fuel required and its cost.

Table 3.10 *Calorific values of fuels*

Fuel	Calorific value
Coal, anthracite	35 MJ/kg
Coke	28 MJ/kg
Oil, domestic heating	45 MJ/kg
Gas, natural	38 MJ/kg
Electricity	3.6 MJ/kWh

Efficiency

Efficiency is a measure of the effectiveness of a system which converts energy from one form to another. The efficiency index or percentage is calculated by comparing the output with the input. The maximum value is 100 per cent although most practical processes are well below the maximum.

$$\frac{\text{Efficiency \%}}{100} = \frac{\text{Useful energy}}{\text{Delivered energy}}$$

or

Delivered energy = Useful energy × 100 / Efficiency %

In general, the useful energy is the output energy from the system which is used to balance the heat losses and heat gains. The delivered energy is the input energy needed for the boiler, or other device, and it is the energy which is paid for.

The overall efficiency of a system depends upon how much of the heat is extracted from the fuel, how much heat is lost through the flue, and

Table 3.11 *Domestic heating efficiencies*

Type of system	House efficiency
Central heating (gas, oil, solid fuel)	60–70 per cent
Gas radiant heater	50–60 per cent
Gas convector heater	60–70 per cent
Electric fire	100 per cent

Note: Although the efficiency of an electrical heating appliance is 100 per cent to the consumer, the overall efficiency of the generation and distribution system is about 30 per cent.

how much heat is lost in the distribution system. The house efficiency is an approximate figure for domestic systems that takes all these effects into account. Some typical values are quoted in table 3.11.

Worked example 3.4

The seasonal heat requirement of a house is 54 GJ, which is to be supplied by a heating system with an overall house efficiency of 67 per cent. The solid fuel used has a calorific value of 31 MJ/kg. Calculate the mass of fuel required for one heating season.

$$\text{Efficiency} = \frac{67}{100}, \text{ Output energy} = 54 \text{ GJ, Input energy} = ?$$

Using

$$\frac{\text{Efficiency \%}}{100} = \frac{\text{Useful energy} \left(\text{system output}\right)}{\text{Delivered energy}\left(\text{boiler input}\right)}$$

$$\frac{67}{100} = \frac{54}{\text{Input energy}}$$

$$\text{Input energy} = 54 \times \frac{100}{67} = 80.597 \text{ GJ}$$

$$= 80597 \text{ MJ}$$

$$\text{Mass of fuel needed} = \frac{\text{Energy required}}{\text{Calorific value}}$$

$$= \frac{80597 \text{ MJ}}{31 \text{MJ/kg}} = 2599.9$$

$$= \textbf{2600 kg}$$

ENERGY REGULATIONS

Many countries have introduced measures which encourage or enforce the efficient use of energy. One reason for the *conservation* of energy is that the Earth's total reserves of convenient fossil fuels, such as oil and gas, are limited. A more recent motive for restricting the use of fuels is an environmental concern about the emission of carbon dioxide produced when fuels are used by transport, by electric power stations, and in buildings.

Energy use in buildings is one of the largest categories of national energy use and therefore any regulations controlling the design of new buildings are an opportunity to make sure that buildings have features which minimise energy use. Good standards of thermal insulation in new buildings, as discussed in the previous chapter, are an important method of controlling heat loss from buildings and good insulation therefore reduces the energy needed to replace that lost heat.

Building Regulations in the United Kingdom originally restricted themselves to the control of thermal insulation in domestic buildings and the early aim was to ensure minimum standards of health and comfort, and to reduce the risk of condensation in housing.

More recent Building Regulations in the United Kingdom have taken as their aim the 'conservation of fuel and power' in all new and converted buildings. The regulations achieve this aim by controlling the following areas of building design and performance:

- Heat loss by transmission through the fabric of the building
- Heat loss by air leakage around openings and through the fabric
- Control systems for space heating and hot water
- Heat loss from vessels and pipes used for hot water
- Heat loss from hot water pipes and hot air ducts used for space heating
- Energy-efficient lighting sources and switching for the lighting.

The following sections give some detail of methods used by United Kingdom Building Regulations for the control of energy in building. The methods are applications of the general theory given earlier in this chapter. The 'philosophy' of how energy use in buildings should be controlled will continue to evolve and you should be alert to future changes. A knowledge of how regulations work in other countries, both near and distant, is a useful way of predicting trends in your own country.

Energy rating, SAP

The overall energy efficiency of a dwelling, such as a house, can be given an *Energy Rating* by using a *Standard Assessment Procedure* (SAP). An SAP Energy Rating of a dwelling is found by using a standard method to calculate the annual energy cost for space heating and water heating in the building. SAP Energy Ratings are expressed on a scale of 0 to 100; the higher the SAP number the better the standard.

Energy Rating	*Energy efficiency
SAP = 0	minimum
SAP = 100	maximum

The Standard Assessment Procedure recommended for use with the Building Regulations for England and Wales takes into account the following factors:

- Thermal insulation of the building
- Efficiency of the heating system
- Controls of the heating system
- Ventilation of the building
- Solar gain by the building
- Price of fuels used for heating spaces and water.

Ratings are adjusted so that they are not affected by differences in the number of people in the dwelling, the floor area, the ownership of particular domestic appliances, and the geographical location of the dwelling. Calculations can be made with the help of worksheets or computer software programs based on the Building Research Establishment Domestic Energy Model, or BREDEM.

Table 3.12 *Typical insulation standards for dwellings*

Element	Standard U-values (W/m^{2}K)	
	SAP 60 or less	over SAP 60
Roofs	0.2	0.25
Flat roof	0.35	0.35
Exposed walls	0.45	0.45
Ground floors/Exposed floors	0.35	0.45
Semi-exposed walls and floors	0.6	0.6
Windows, doors and rooflights	3.0	3.3
Maximum area double glazing	22.5% floor area	

Notes
The SAP (Standard Assessment Procedure) Energy Rating of a dwelling is based on annual energy costs.
Areas of walls, roofs and floors are measured between internal faces.
Exposed elements are those exposed to outside air (including a suspended floor) or an element in contact with the ground.
Semi-exposed elements are those separating a heated space from an unheated space which does not meet standards.
The values are summarised from typical Building Regulations for England and Wales.
Such values are revised periodically and the latest regulations applicable to a particular case should be consulted.

Insulation of the building fabric

To allow design flexibility, the regulations allow fabric insulation requirements to be met by several different approaches, as summarised below for dwellings:

- **Elemental method**: The matching or bettering of standard U-values for individual construction elements. The calculation method takes account of *thermal bridges*. Dwellings with higher (better) SAP ratings can use slightly higher (inferior) U-value targets.

- **Target *U*-value method**: Provides a maximum *U*-value for the exposed fabric as a whole and can allow for solar gain and more efficient heating systems if desired. Dwellings with higher (better) SAP ratings can use higher (inferior) *U*-value targets.
- **Energy Rating method**: Allows buildings with high calculated SAP Energy Ratings to satisfy regulations, subject to limitations on thermal bridging and condensation.

Buildings other than dwellings can also use the elemental method, described above, to match or to better a set of standard *U*-values for such buildings. In addition, buildings other than dwellings can show compliance with the regulations by using the following methods:

- **Calculation method**: Allows greater flexibility in areas of fabric elements, windows and doors. Calculations need to show that the rate of heat loss for the proposed building is not greater than the total rate of heat loss from a notional building of the same size and shape designed to comply with the Elemental method.
- **Energy use method**: Allows a free design of the building using valid methods of energy conservation. Calculations need to show that the annual energy use of the proposed building is less than the calculated energy use of a similar building designed to comply with the Elemental method.

Thermal bridging around openings

A thermal bridge is a portion of a structure, usually a junction, where lower insulation allows higher heat loss and therefore lowers the overall thermal insulation of the structure. Examples are show in figure 2.4 in the previous chapter. The Building Regulations require that thermal bridging around windows doors and other openings is limited by adopting the following design details or their equivalent.

Area of thermal bridging	Measures required
Lintels and Jambs and Sills	Insulating blockwork or Internal insulation on the inside face of the wall and its return or Partial cavity fill in the wall or Full cavity fill in the wall

Infiltration

Cold air from outside can infiltrate into a building by various leakage paths in the building fabric. This cold air, which is not part of any ventilation system, has to be heated and therefore wastes energy. The Building Regula-

Table 3.13 *Typical insulation standards for buildings other than dwellings*

Type of building	Element	Standard *U*-values (W/m²K)
Buildings other than dwellings	Roofs	0.25
	Flat roof	0.35 for residential
		0.45 for other buildings
	Exposed walls	0.45
	Ground floors/ Exposed floors	0.45
	Semi-exposed walls and floors	0.6
	Windows, doors and rooflights	3.3

Type of building	Element	Maximum area of double glazing
Residential	Windows and doors	30% exposed wall area
Places of assembly, offices and shops	Windows and doors	40% exposed wall area
Industrial and storage buildings	Windows and doors	15% exposed wall area
All buildings other than dwellings	Rooflights	20% roof area

Notes
Areas of walls, roofs and floors are measured between internal faces.
Exposed elements are those exposed to outside air (including a suspended floor) or an element in contact with are ground.
Semi-exposed elements are those separating a heated space from an unheated space which does not meet standards.
The values are summarised from typical Building Regulations for England and Wales.
Such values are revised periodically and the latest regulations applicable to a particular case should be consulted.

tions require that air leakage through the building fabric is limited, as far as is practicable, by the following types of preventative measure.

Area of infiltration	Measures required
Dry lining on masonry walls	Sealing of gaps with windows and doors sealing of gaps at junctions with wall, floors and ceilings
Timberframe construction	Complete sealing of vapour control membrane
Windows, doors, rooflights	Fitting draught-stripping in frames of openable elements
Loft hatches	Sealing around the hatch
Services pipes	Sealing at around boxing at floor and ceiling levels Sealing around pipes which project or penetrate into voids

Space heating controls

Space heating or central heating in houses and flats is commonly provided by boiler-hot water systems, electric storage heaters, and electric panel heaters. The Building Regulations require certain features provision and those for dwellings are summarised below.

Feature required	Aim
Zone controls, such as room thermostats, thermostatic radiator valves	to control the temperatures independently in areas which require different temperatures, such as living and sleeping areas
Timing controls	to control the periods when heating systems operate
Boiler control interlocks	to switch boiler off when no heat is required for boiler to fire only when thermostat is calling for heat to prevent unnecessary boiler cycling

For buildings other than dwellings, the requirements for space heating controls have similar aims to those listed above but the requirements are specified in terms suited to larger installations.

Hot water controls and insulation

The Building Regulations require that hot water storage systems have suitable features which are summarised below.

Feature required	Aim
Heat exchanger of appropriate capacity	to allow effective control
Thermostat	to shut off the supply of heat when the storage temperature is reached to switch off the boiler when no heat is required in the case of a central heating system
Timer, which may be part of central heating system	to shut off the supply of heat when water heating is not required

The requirements for the insulation of services are summarised below.

Feature	Measure required
Hot water vessels	Insulation that meets published standards such as BS1566.
Unvented hot water systems	Additional insulation to control heat loss through the safety fittings and pipework
Pipes and ducts, which do not contribute to useful heat requirement	Insulation that meets published standards
Hot pipes connected to hot water storage vessels	Insulation for at least one metre from points of contact or up to the point where they become concealed.

Alternative approaches, using published standards, are permitted in order to meet the requirements for space heating controls and hot water storage controls.

Lighting

For buildings other than dwellings which have more than $100\,m^2$ of floor area with artificial lighting the Building Regulations require the features summarised below. Lighting terms and equipment are explained in chapter 6.

Feature	Measure required
Lamps of high efficiency	types such as high pressure sodium, metal halide, induction lighting, tubular fluorescent, compact fluorescent
Lighting controls, such as local switching, time switching	to encourage the maximum use of daylight without endangering the passage of occupants

Exercises

1 The external wall of a room measures 4.8 m by 2.6 m and has an average *U*-value of $1.8\,W/m^2\,K$. The internal air temperature is 21 °C, the mean radiant temperature is 18 °C, and the external air temperature is 0 °C.
   (a) Calculate the environmental temperature inside the room.
   (b) Use the environmental temperature to calculate the rate of heat loss through the wall.

2 A house has a floor area of $92\,m^2$ and a ceiling height of 2.5 m. The average inside air temperature is kept at 18 °C, the outside air temperature

is 6 °C, and the average infiltration rate is 1.5 air changes per hour. The volumetric specific heat capacity of the air is 1300 J/m³ K.

(a) Calculate the rate of ventilation heat loss.

(b) Calculate the cost of the heat energy lost during 24 hours if the above conditions are maintained and replacement heat costs 5 pence per megajoule.

3 A sports pavilion has internal dimensions of 11 m × 4 m × 3 m high. 20 per cent of the wall area is glazed and the doors have a total area of 6 m². The U-values in W/m K are: walls 1.6, windows 5.5, doors 2.5, roof 1.5, floor 0.8. The inside air temperature is maintained at 18 °C when the outside air temperature is −2 °C. There are four air changes per hour and the volumetric specific heat capacity of air is 1300 J/m³ K. The heat gains total 2200 W.

(a) Calculate the net rate of heat loss from this building.

(b) Calculate the surface area of the radiators required to maintain the internal temperature under the above conditions. The output of the radiators is 440 W/m² of radiating surface area.

4 A room has 7.5 m² area of single glazed windows, which have a U-value of 5.6 W/m² K. It is proposed to double glaze the windows and reduce the U-value to 3.0 W/m² °C. During a 33 week heating season the average temperature difference across the windows is 7 °C.

(a) For both types of glazing, calculate the total heat lost during the heating season.

(b) Obtain current figures for the cost of electrical energy and the approximate cost of double glazing such windows. Estimate the number of years required for the annual fuel saving to pay for the cost of the double glazing.

5 The average rates of heat loss for a particular house are 1580 W total fabric loss and 870 W ventilation loss. The seasonal heat gains of the house total 27 500 MJ. The fuel used has a calorific value of 32 MJ/kg and the heating system has an overall efficiency of 75 per cent.

(a) Calculate the input heat required during a heating season of 33 weeks.

(b) Calculate the mass of fuel required to supply one season's heating.

Answers

1 (a) 19 °C; **(b)** 427 W

2 (a) 1495 W; **(b)** 646 pence

3 (a) 8030 W; **(b)** 18.25 m²

4 (a) 5.868 GJ; **(b)** 3.144 GJ

5 (a) 21.398 GJ; **(b)** 892 kg

4 Ventilation, Humidity and Condensation

This chapter is concerned with ventilation and with moisture in the air, and how they relate to human comfort and to the performance of buildings. These topics are important in the thermal design and behaviour of buildings and are governed by the properties and principles of heat, gases, and vapours described in chapter 1.

VENTILATION

Ventilation in buildings is the process of changing the air in a room or other internal space. This process should be continuous with new air taken from a clean source. Although we require oxygen for life, a build-up of carbon dioxide is more life-threatening and a general build-up of odours will be more critical long before there is danger to life. In addition to the comfort of the occupants the ventilation of a building has other objectives, such as those included below:

- Supply of oxygen
- Removal of carbon dioxide
- Control of humidity for human comfort
- Control of air velocity for human comfort
- Removal of odours
- Removal of micro-organisms, mites, moulds, fungi
- Removal of heat
- Removal of water vapour to help prevent condensation
- Removal of particles such as smoke and dust
- Removal of organic vapours from sources such as cleaning solvents, furniture, and building products
- Removal of combustion products from heating and cooking
- Removal of ozone gas from photocopiers and laser printers
- Removal of methane gas and decay products from ground conditions.

The 'old' air being replaced has often been heated and, to conserve energy, the rate of ventilation may be limited or heat may be recovered from the extracted air.

87

Other factors to consider in the provision of ventilation include the following:

- Control of fire
- Conservation of energy
- Noise from the system.

Ventilation requirements

Adequate ventilation of buildings and spaces is required by several sources of regulations in Britain such as the Building Regulations, Workplace Regulations, Housing Acts, and the Health and Safety at Work legislation.

Various authorities produce technical guidance for providing adequate ventilation, and important sources include British Standards, Building Research Establishment (BRE) and the Chartered Institution of Building Services Engineers (CIBSE).

Table 4.1 *Typical ventilation rates*

Element	Rate
Commercial kitchens	20–40 air changes per hour
Restaurants (with smoking)	10–15 air changes per hour
Classrooms	3–4 air changes per hour
Offices	2–6 air changes per hour
Domestic rooms	1 air changes per hour
General occupied rooms	8 litres/second fresh air per occupant

The method of specifying the quantity and rate of air renewal varies and includes the following:

- Air supply to a space – such as 1.5 room air changes per hour
- Air supply to a person – such as 8 litres per second per person.

In the simple case it can be assumed that if air is actively extracted, such as by a fan, then air will flow in to replace the extracted air. So the supply rate will match the extract rate, although the source of the supply may need to be considered in the ventilation design.

Other methods of ensuring minimum ventilation, as used by Building Regulations, include defining the following:

- Number or area of open windows to provide *rapid ventilation*
- Area of opening to provide *trickle ventilation*
- Performance of fan or stack to provide *extract ventilation*.

Table 4.2 summarises typical ventilation requirements for dwellings.

Air supply rates
Useful conversion between systems of specifying ventilation are given in the table 4.3 and worked example 4.1.

Table 4.2 *Typical ventilation standards for dwellings*

Type of room	Rapid ventilation e.g. opening windows	Background ventilation e.g. trickle vents	Extract ventilation e.g. fan rates
Habitable room	$1/20^{th}$ of floor area	$8000 \, mm^2$	–
Kitchen	opening window (no minimum size)	$4000 \, mm^2$	60 litres/second or PSV*
Utility room	opening window (no minimum size)	$4000 \, mm^2$	30 litres/second or PSV*
Bathroom (with or without WC)	opening window (no minimum size)	$4000 \, mm^2$	15 litres/second or PSV*
WC (separate from bathroom)	$1/20^{th}$ of floor area or mechanical extract at 6 litres/second	$4000 \, mm^2$	–

* PSV = Passive Stack Ventilation, which relies on the natural stack effect caused by temperature differences.

Table 4.3 *Useful ventilation rate conversions*

1000 litres = $1 \, m^3$
1 litre/s = 3600 litres/hour
1 litre/s = $3.6 \, m^3$/hour

For example, a common standard per person is
8 litres/s = $30 \, m^3$/hour

Worked example 4.1

An extract fan is required for a room with sanitary accommodation (WC or urinal) with the following details:

target air change per hour = 3 ach
volume of room $2 \, m \times 3 \, m \times 2.5 \, m = 15 \, m^3$

So

3 ach = $3 \times 15 = 45 \, m^3$/hour = 45 000 litres/hour
45 000 litres/hour = $45\,000/60 \times 60 = 15.5$ litres/second

Therefore the extract fan needs to have an extraction rate of at least **15.5 litres/second.**

Ventilation installations

The design of ventilation systems needs to take account of the following factors:

- Volume of air
- Movement of air
- Distribution of air
- Filtration 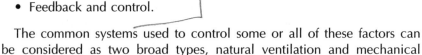 — mention
- Temperature change
- Humidity change
- Energy conservation
- Feedback and control.

The common systems used to control some or all of these factors can be considered as two broad types, natural ventilation and mechanical ventilation.

Natural ventilation

Natural ventilation is provided by two broad mechanisms:

- **Air pressure differences:** as caused by the wind direction or movement over and around a building
- **Stack effect:** caused by the natural convection of warm air rising within a building.

A simple example of using air pressure differences is opening windows on either side of a building. See also the section on Fluid Flow in chapter 12 for more information about the effects of pressure and velocity in fluids such as air.

The stack effect is particularly noticeable when there is a considerable difference in height, such as in the stairwell of a tall building. This effect in a stairwell is a nuisance and a fire danger in a stairwell but it can be made good use of in a ventilation system. Some modern environmentally-friendly buildings use the stack effect to avoid the need for mechanical ventilation.

Mechanical ventilation

The use of mechanical ventilation make it possible to use spaces, such as deep within buildings, that could not be easily ventilated by natural means. The mechanical fans depend on a supply of energy, usually electricity.

A simple extract or input mechanical system provides movement of air but only a limited degree of control. A *plenum* system gives better control of air by using ductwork and usually has the ability to heat the air if

Table 4.4 *Features of typical mechanical ventilation installation*

Installation feature	Purpose and features
Fresh air intake	must be carefully situated
Recirculated air	can be mixed with fresh air
Air filter	removes particles and contaminants
Heater	elements or coils over which air passes
Fan	provides air movement for intake and distribution
Ductwork	guides distribution of air
Diffusers	control distribution of air into room

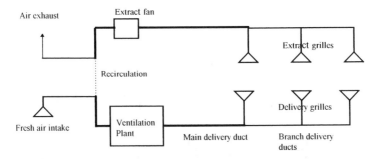

Figure 4.1 *Schematic diagram of typical ventilation system*

necessary. The features of a typical system are shown in table 4.4 and in figure 4.1.

Air conditioning

An air conditioning installation has the aim of producing and maintaining a designed internal air environment, despite the variations in external air conditions. The equipment therefore has to be able to heat and cool the air, to humidify and dehumidify the air, and to respond automatically to changes in the external air.

Typical air conditioning equipment is described in table 4.5 and in figure 4.2. The removal of water from the air, to lower the humidity, is usually achieved by cooling the air with refrigeration plant and then reheating the air before passing it into the distribution system. More information about changes in humidity is given in the next section of this chapter, but the feature to note is that air conditioning plant uses significant amounts of energy.

Table 4.5 *Features of typical air conditioning installation*

Installation feature	Purpose and features
Fresh air intake	must be carefully situated
Recirculated air	can be mixed with fresh air
Air filter	removes particles and contaminants
Heater	elements or coils over which air passes
Preheater	to provide initial heat energy if needed
Chiller	lowers temperature and therefore removes moisture
Humidifier	adds moisture to the air if needed
Reheater	adjusts final temperature of air if necessary
Fan	provides air movement for intake and distribution
Ductwork	guides distribution of air
Diffusers	control distribution of air into room

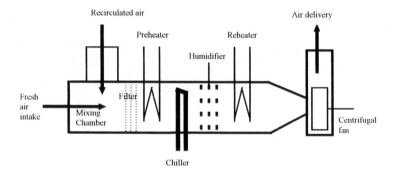

Figure 4.2 *Schematic diagram of air conditioning plant*

HUMIDITY

Humidity is the study of moisture in the atmosphere. The maximum proportion of water vapour in air is about 5 per cent by weight yet this relatively small amount of moisture produces considerable effects. Human comfort, condensation in buildings, weather conditions, and water supplies are important environmental topics dependent on humidity. The amount of moisture in the air also influences the durability of materials, the drying of materials, the operation of industrial processes, and the growth of plants.

Most of the moisture in the atmosphere is a result of evaporation from the sea, which covers more than two-thirds of the Earth's surface. At any particular place, natural humidity is dependent on local weather conditions and, inside a building, humidity is further affected by the thermal properties and the use of the building.

Water vapour

A vapour can be defined as a substance in the gas state which may be liquefied by compression. Water vapour, for example, is formed naturally in the space above liquid water which is left open to the air. This process of *evaporation* occurs because some liquid molecules gain enough energy, from chance collisions with other molecules, to escape from the liquid surface and become gas molecules. The latent heat required for this change of state is taken from the other molecules of the liquid, which therefore becomes cooled. The rate of evaporation increases if there is a movement of air above the liquid.

- **Water vapour is invisible.**

Steam and mist, which can be seen, are actually suspended droplets of water liquid, not water vapour. The molecules of water vapour rapidly occupy any given space and exert a vapour pressure on the sides of any surface that they are in contact with. This pressure behaves independently of the other gases in the air – an example of Dalton's Law of Partial Pressures described in chapter 1.

Saturation

If the air space above a liquid is enclosed then the evaporated vapour molecules collect in the space and the vapour pressure increases. Some molecules are continually returning to the liquid state and eventually the number of molecules evaporating is equal to the number of molecules condensing. The air in the space is then said to be saturated. *Saturated air* is a sample which contains the maximum amount of water vapour possible at that temperature.

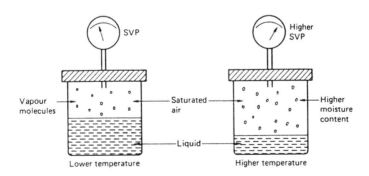

Figure 4.3 *Saturated air*

Vapour pressure increases as the amount of water vapour increases and at saturation the vapour pressure reaches a steady value called the saturated vapour pressure.

- *SATURATED VAPOUR PRESSURE (SVP)* **is the vapour pressure of the water vapour in an air sample that contains the maximum amount of vapour possible at that temperature**

Saturated vapour pressure is found to increase with increase in temperature. Table 4.6 lists the values for the saturated vapour pressure of water vapour at different temperatures.

Increased vapour pressure indicates an increased moisture content, therefore the saturation of a fixed sample of air is delayed if the temperature of the air is raised. This property gives rise to the important general principle:

- Warm air can hold more moisture than cold air.

If an unsaturated sample of moist air is cooled then, at a certain temperature, the sample will become saturated. If the sample is further cooled below this dew point temperature then some of the water vapour must condense to liquid. This condensation may occur on surfaces, inside materials, or around dust particles in the form of cloud or fog.

Specification of humidity

Some of the properties of water vapour can be used to specify the amount of moisture in a sample of air. The different variables and their applications are listed below and described in the sections that follow.

Table 4.6 *Saturated vapour pressures of water vapour*

Temperature (°C)	SVP (Pa)	Temperature (°C)	SVP (Pa)
0	610	13	1497
1	657	14	1598
2	705	15	1704
3	758	16	1818
4	813	17	1937
5	872	18	2063
6	935	19	2197
7	1001	20	2337
8	1072	25	3166
9	1148	30	4242
10	1227	40	7375
11	1312	50	12340
12	1402	100	101325

- Moisture content
- Vapour pressure
- Dew point
- Relative humidity.

Moisture content

Moisture content is a measure of absolute humidity – the actual quantity of water vapour present in the air.

$$\textbf{Moisture content} = \frac{\text{mass of water vapour in air sample}}{\text{mass of that air sample when dry}}$$

UNIT: kg/kg of dry air (or g/kg)

Moisture content is not usually measured directly but it can be obtained from other types of measurement. Moisture content values are needed, for example, in determining what quantity of water an air conditioning plant needs to add or to extract from a sample of air.

Vapour pressure

The molecules of water vapour in the air exert a pressure that increases as the amount of water vapour increases.

- **VAPOUR PRESSURE is the partial pressure exerted by the molecules of a vapour**

  UNIT: pascal (Pa)

Other, non-standard units which may still be found in use include:

- millibars (mb), where 1 mb = 101 Pa
- mm of mercury, where 1 mm = 133 Pa.

The vapour pressure of moisture in the air behaves independently of the other gases in the air. Vapour pressure values are usually derived from other measurements and one of the uses of vapour pressure is to determine the rate at which water vapour moves through materials.

Dew point

If moist air is cooled then, at a certain temperature, the air becomes saturated with water vapour. If this air is in contact with a surface that is at or below this temperature then a thin film of liquid will form. This effects is known as dew or condensation.

- **The DEW POINT is the temperature at which a fixed sample of air becomes saturated**

  UNIT: °C or K

A sample of air with a fixed moisture content has a constant dew point, even if the air temperature changes. The dew point is particularly relevant in the study of condensation in buildings. Dew point values can be measured directly or derived from other measurements.

Relative humidity

The relative humidity (RH) of a sample of air compares the actual amount of moisture in the air with the maximum amount of moisture the air can contain at that temperature. The strict definition of relative humidity is given below:

$$\textbf{Relative humidity} = \frac{\text{vapour pressure of sample}}{\text{SVP of sample at same temp.}} \times 100$$

UNIT: per cent RH, at a specified temperature

It is also common practice to describe humidity in terms of *percentage saturation*, which is defined by the following formula:

$$\textbf{\% Saturation} = \frac{\text{mass of water vapour in air sample}}{\substack{\text{mass of water vapour required to} \\ \text{saturate sample at same temperature}}} \times 100$$

Percentage saturation and relative humidity are identical only when air is perfectly dry (0 %) or fully saturated (100 %). For temperatures in the range 0 to 25 °C the difference between relative humidity and percentage saturation is small.

A convenient alternative formula for relative humidity is given below:

$$\textbf{Relative humidity} = \frac{\text{SVP at dew point}}{\text{SVP at room temperature}}$$

An RH of 100 per cent represents fully saturated air, such as occurs in condensation on a cold surface or in a mist or fog. An RH of 0 per cent represents perfectly dry air; such a condition may be approached in some desert conditions and in sub-zero temperatures when water is frozen solid.

The SVP or saturated moisture content varies with temperature; therefore RH changes as the temperature of the air changes. Despite this dependence on temperature, RH values are a good measurement of how humidity affects human comfort and drying processes.

Because warm air can hold more moisture than cool air, raising the temperature increases the SVP or the saturation moisture content. The denominator on the bottom line of the RH fraction then increases and so the RH value decreases. This property gives rise to the following general effects for any sample of air:

- Heating the air lowers the relative humidity
- Cooling the air increases the relative humidity.

Worked example 4.2

A sample of air has an RH of 40 per cent at a temperature of 20 °C. Calculate the vapour pressure of the air (given: SVP of water vapour = 2340 Pa at 20 °C).

RH = 40 per cent, vp = ? SVP = 2340 Pa

Using

$$RH = \frac{VP}{SVP} \times 100$$

$$40 = \frac{VP}{2340} \times 100$$

$$vp = \frac{2340 \times 40}{100} = 936$$

Vapour pressure = **936 Pa**

Hygrometers

Hygrometry, also called psychrometry, is the measurement of humidity. The absolute humidity (moisture content) of a sample of air could be measured by carefully weighing and drying it in a laboratory. Usually however, other properties of the air are measured and these properties are then used to calculate the value of the RH.

- Hygrometers, or *psychrometers,* are instruments which measure the humidity of air.

Hair, paper hygrometer
The hair hygrometer and the paper hygrometer make use of the fact that animal hair or paper change their dimensions with changes in moisture content. These instruments can be made to give a direct reading of RH on their dials, but they need calibration against another instrument.

Dew point hygrometer
The temperature of the dew point is a property that can be observed and measured by cooling a surface until water vapour condenses upon it. The dew point temperature and the room temperature can then be used to obtain an RH value from tables or charts.

The *Regnault hygrometer* is one form of dew point hygrometer. A gentle passage of air is bubbled through liquid ether in a container and causes the ether to evaporate. The latent heat required for this evaporation is taken from the liquid ether and from the air immediately surrounding the container, which therefore becomes cooled. The polished outside surface of the container is observed and the moment a thin film of mist appears the temperature is noted from the thermometer in the ether. When the instrument is used carefully, this observed temperature is also the dew point temperature of the surrounding air.

Wet-and-dry-bulb hygrometer

If the bulb of a thermometer is wrapped in a wetted fabric the evaporation of the moisture absorbs latent heat and will cool the bulb. This wet bulb thermometer will record a lower temperature than an adjacent dry bulb thermometer (an ordinary thermometer). Dry air with a low RH causes rapid evaporation and produces a greater wet bulb depression than moist air. Saturated air at 100 per cent RH causes no net evaporation and the dry bulb and wet bulb thermometers then record the same temperature. This changing difference between the wet and the dry thermometers – the 'depression' – can therefore be used as an indicator of relative humidity.

The two thermometers are mounted side by side, and a water supply and

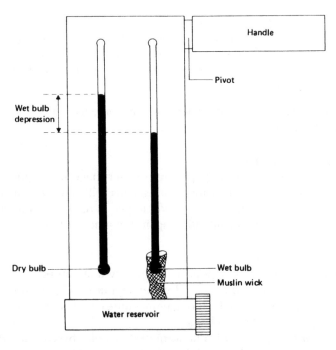

Figure 4.4 *Whirling sling hygrometer*

fabric wick are connected to the wet bulb. The *whirling sling hygrometer*, shown in figure 4.4, is rapidly rotated by means of a handle before reading, to circulate fresh air around the bulbs. Each type of hygrometer has calibrated tables which give the RH value for a particular pair of wet and dry bulb temperatures.

Psychrometric chart

The different variables used to specify the amount of water vapour in the air are related to one another. These relationships between different types of measurement can be expressed in the form of tables of values, or in the form of graphs.

- **A *PSYCHROMETRIC CHART* is a set of graphs which are combined so that they plot the relationships between the different variables used to specify humidity**

Figure 4.5 is one form of psychrometric chart, which displays the following measurements:

- **Dry-bulb temperature:** the ordinary air temperature, is read from a horizontal scale on the base line.
- **Moisture content:** the mixing ratio, is read from one of the right-hand vertical scales.
- **Vapour pressure:** read from one of the right-hand vertical scales.
- **Wet-bulb temperature:** read from the sloping straight lines running from the saturation line.
- **Dew point temperature:** read from the horizontal lines running from the saturation line.
- **Relative humidity:** read from the series of curves running from the left-hand vertical scale.

Notice that the saturation curve represents 100 per cent RH and that, at this saturated condition, the dry-bulb, the wet-bulb and the dew point temperatures all have the same value.

A psychrometric chart is strictly valid for one value of atmospheric pressure, the sea level pressure of 101.3 kPa being a common standard. Some versions of the psychrometric chart contain additional information, such as *sensible heat* and *latent heat* contents, which are of particular use to heating engineers.

Use of the psychrometric chart

The temperature and moisture content of a particular sample of air is described by a pair of values. These two values also act as coordinates on the chart and define a single *state point* for the air. Moving to any other state point on the chart represents a change in the condition of the air.

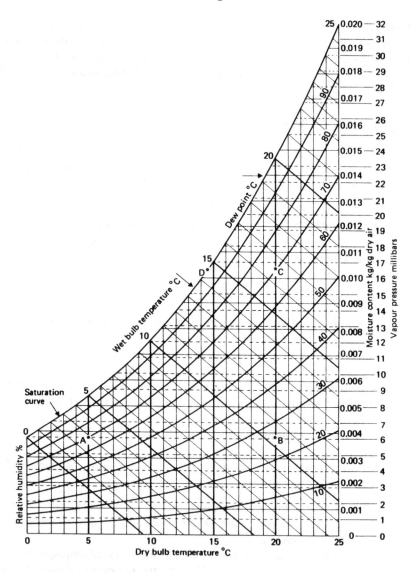

Figure 4.5 *Psychrometric chart of moisture and temperature*

The direction of movement on the chart also represents specific types of air conditioning change. For example, horizontal movements mean that the moisture content of the air remains constant, as happens when air is simply heated without humidification. The relative humidity, however, does change during horizontal movements.

The following air conditions and changes are marked on figure 4.5:

Point A: Represents air of 5 °C and 70 per cent RH. From the scales it can also be read that the moisture content is 0.0038 kg/kg and the vapour pressure is 6 mb.

Point B: Represents air where the dry bulb has been raised to 20 °C and so the RH has decreased to 25 per cent. The moisture content and the vapour pressure are unchanged.

Point C: Represents air at the same dry bulb but with an increase in moisture content to 0.103 kg/kg. The vapour pressure has also increased. The RH has been restored to 70 per cent. The wet bulb at this state is 16.5 °C.

Point D: Represents the dew point of this same air at 14.5 °C. This is the temperature at which the air is saturated and has 100 per cent RH.

Worked example 4.3

External air at 0 °C and 80 per cent RH is heated to 18 °C. Use the psychrometric chart to determine the following information:

(a) the RH of the heated air;
(b) the RH of the heated air if 0.005 kg/kg of moisture is added;
(c) the temperature at which this moistened air would first condense.

Initial conditions: dry bulb = 0 °C, RH = 80 per cent, so moisture content = 0.003 kg/kg

(a) For the heated air
    moisture content = 0.003 kg/kg
         dry bulb = 18 °C
So, reading from chart
         RH = **23 per cent**

(b) For the moistened air
    moisture content = 0.003 + 0.005 = 0.008 kg/kg
         dry bulb = 18 °C
so, reading from chart
         RH = **62 per cent**

(c) For condensation
         RH = 100 per cent
    moisture content = 0.008
So, reading from chart
         dew point = **10.8 °C** gives condensation

CONDENSATION IN BUILDINGS

Condensation in buildings is a form of dampness caused by water vapour in the air. Among the effects of condensation are misting of windows, beads of water on non-absorbent surfaces, dampness of absorbent materials, and mould growth.

Condensation is not likely to be a problem in a building where it has been anticipated and designed for, as in a tiled bathroom or at an indoor swimming pool. Unwanted condensation, however, is a problem when it causes unhealthy living conditions; damage to materials, to structures and to decorations; or general concern to people.

Condensation as a problem is a relatively recent concern and one that has been increasing. It is affected by the design of modern buildings and by the way in which buildings are heated, ventilated, and occupied. These factors are considered in the sections on the causes of condensation and the remedies for it.

Principles of condensation

Warm air can hold more moisture than cold air. If air in a building acquires additional moisture, this increased moisture content will not be seen in places where the air is also warmed. But if this moist air comes into contact with colder air, or with a cold surface, then the air is likely to be cooled to its dew point. At this temperature the sample of air becomes saturated, it can no longer contain the same amount of water vapour as before, and the excess water vapour condenses to liquid.

- **Condensation in buildings occurs whenever warm moist air meets surfaces that are at or are below the dew point of that air**

It is convenient to classify the effects of condensation into two main types:

- Surface condensation
- Interstitial condensation.

Surface condensation
Surface condensation occurs on the surfaces of the walls, windows, ceilings, and floors. The condensation appears as a film of moisture or as beads of water on the surface and is most obvious on the harder, more impervious surfaces. An absorbent surface may not show condensation at first, although persistent condensation will eventually cause dampness.

Interstitial condensation
Interstitial condensation occurs within the construction of a building as indicated in figure 4.6. Most building materials are, to some extent, perme-

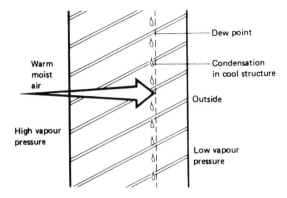

Figure 4.6 *Interstitial condensation*

able to water vapour (that is, they allow the passage of air containing moisture). If this air cools as it passes through the structure then, at the dew point temperature, condensation will begin to occur inside the structure. The dampness caused by this interstitial condensation can damage important structural materials, such as steelwork, and can make insulating materials less effective.

Causes of condensation

The general requirements for condensation are moist air and cold surfaces. The factors that influence the production of condensation can be considered under the following headings:

- Moisture sources
- Air temperatures
- Structural temperatures
- Ventilation
- Use of buildings.

Moisture sources
The moisture present in any sample of air comes from a source of water. Local weather conditions determine the initial moisture state of the air but if the outside air is cool then it has a low moisture content and does not cause condensation when it is inside a building. Warm humid weather can cause condensation, especially if there is a sudden change from cool weather to warm moist conditions and the room surfaces are slow to heat up. In temperate climates like that of the UK, the amount of rainfall does not have a big influence on condensation inside buildings.

• Condensation is **not** usually caused by damp weather.

Most of the moisture content of the air inside dwellings comes from the occupants and their activities. An average family produces 10 to 20 kg of moisture per day by the activities of normal breathing, cooking, and washing. Flueless devices, such as gas cookers or bottle gas heaters, release great quantities of water vapour and increase the risk of condensation. The moisture produced in kitchens and bathrooms can quickly diffuse into other rooms and cause condensation far away from the original source of the moisture.

The construction processes involved in a traditionally-built brick house add about 5000 kg of water to the structure – mainly by the water used for mixing concrete, mortar and plaster, and by exposure to weather. This water dries out during the first months of occupancy and adds to the moisture content of the air inside the building.

Temperatures

Warm air can hold more moisture than cool air and is therefore less likely to cause condensation, but the extra moisture contained in the warm air can cause condensation when that air is cooled. For example, the kitchen itself may be warm enough to be free from condensation but the same moisturised air may cause condensation when it diffuses to a cooler part of the house.

Condensation occurs on those surfaces, or in those materials, that have a temperature at or below the dew point of the moist air. If moisture is progressively added to the air in a room then condensation will first occur on the coldest surfaces, on windows and cold pipes for example. The temperatures of room surfaces are lowest where the thermal insulation is lowest. Condensation can indicate the areas of worst insulation, particularly those caused by 'cold bridges' such as on lintels above windows.

The speed at which structures change their temperature affects condensation. Heavyweight construction, with a high thermal capacity, will be slower to respond to heating than a lightweight construction. A brick wall or a concrete floor, for example, is slow to increase its temperature when heated and condensation may occur for a period while it is heating. The positioning of insulation on the inside of heavyweight structures will allow the surface temperatures to rise more rapidly.

Ventilation

The air outside a building usually has a much lower moisture content than the air inside, because cool air cannot hold as much moisture as warm air. Ventilating a building therefore lowers the moisture content inside and reduces the risk of condensation. It is theoretically possible to avoid all condensation by adequate ventilation but as the ventilation rate increases the heat loss in the discarded air also increases.

The natural ventilation rates in dwellings have been decreasing over the years and this fact greatly contributes to the incidence of condensation in buildings. Chimneys have been eliminated in many modern houses and blocked-up in older houses. The construction of modern doors, windows, and floors usually provides better seals against the entry of outside air than in the past. As a result, the natural ventilation in dwellings has fallen from typical rates of 4 air changes per hour to less than 1 air change per hour.

Even with higher ventilation rates it is possible to have stagnant volumes of moist air where condensation still occurs; behind furniture and inside wardrobes, for example.

Use of buildings

Over the years there have been changes in the design of buildings and changes in the way that we live in these buildings. These changes have tended to increase the risk of condensation.

Methods of heating buildings have changed and, at the same time, people expect higher standards of thermal comfort and cleanliness than they did in the past. The presence of chimneys in older houses aided natural ventilation and the use of a fire increased the ventilation by drawing air for combustion. The cool draughts that resulted were offset to some extent by the direct warming effect of radiant heat from the fire.

Thermal comfort by means of warm air from modern convective heating requires low levels of air movement and this leads to the draught-proofing of houses. Public campaigns for better thermal insulation also encourage lower ventilation rates and these measures can increase the risk of condensation unless care is taken.

Moisture-making activities such as personal bathing and washing laundry have increased because of improved facilities and changing social customs. The laundry is often dried inside the building, sometimes by a tumble drier which is not vented to the outside.

Many dwellings are only occupied during the evening and the night. This pattern of living tends to compress the cooking and washing activities of a household into a short period, at a time when windows are unlikely to be opened. During the day, when the dwelling is unoccupied, the windows may be left closed for security reasons and heating is turned off, allowing the structure to cool.

Remedies for condensation

The preceding discussions of the causes of condensation also reveal methods for helping to prevent condensation. These various preventive measures can be summarised as three main types of remedy for persistent condensation inside buildings:

- Ventilation
- Heating
- Insulation.

A combination of these three remedies is usually necessary. The use of vapour barriers as a measure against interstitial condensation is discussed under a separate heading. *Anti-condensation* paint can be useful for absorbing temporary condensation but it is not a permanent remedy for condensation.

Ventilation
Ventilation helps to remove the moist air, which might otherwise condense if it is cooled inside the building. The ventilation is most effective if it is used near the source of moisture (for example, an extractor fan in the kitchen). Care should be taken that natural ventilation from windows does not blow steam into other rooms; kitchen and bathroom doors should be self-closing.

Some windows in each room should have a ventilator, which can give a small controlled rate of ventilation without loss of security, loss of comfort, or loss of significant amounts of heat energy. The cost of the heat lost in necessary ventilation is small and it should decrease as techniques of heat recovery from exhaust air are applied to houses.

Heating
Heating a building raises the temperature of the room surfaces and helps keep them above the dew point of the air inside the building. Heated air also has the ability to hold more moisture, which can then be removed by ventilation before it has a chance to condense upon a cold surface.

The level and timing of the heating used in a building affect condensation and, in general, long periods off low heating are better than short periods at high temperature. Heavyweight construction, such as concrete and brickwork, should not be allowed to cool completely, and a continuous level of background heating helps maintain temperatures. Heating devices without flues should not be used if there is a risk of condensation. A paraffin heater, for example, gives off about 1 litre of water for each litre of fuel used.

Insulation
Thermal insulation reduces the rate at which heat is lost through a structure and will help in keeping inside surfaces warm, although insulation by itself cannot keep a room warm if no heat is supplied. Insulation placed on the inside surface of a heavyweight construction, such as a brick wall, helps to raise the surface temperature more quickly when the room is heated. However, with the insulation near the inside of a room, the outer part of the wall

will remain cool and a vapour barrier is needed to prevent interstitial condensation, as discussed in the next section.

CONDENSATION CONDITIONS

The risk of condensation occurring on or in a building material depends upon the temperature and the humidity of the air on both sides of the structure and also upon the resistance of the material to the passage of heat and vapour. The thermal resistance of materials was used to calculate *U*-values in chapter 2. This section introduces the similar idea of vapour resistance and uses it to calculate when and where condensation may occur in a structure.

Vapour transfer

The vapour pressure of a sample of air increases when the moisture content increases. Because the occupants of a building and their activities add moisture to the air, the vapour pressure of the inside air is usually greater than that of the cooler outside air. This pressure difference results in the following general rule:

- Water vapour passes through structures from inside to outside.

The rate at which water vapour passes through a structure depends upon the vapour *permeability* of the various building materials present – that is, the ease with which they permit the diffusion of water vapour. This property of a material concerning the behaviour of water vapour is often totally different from the behaviour of that same material concerning water liquid. It is relatively easy to make a material waterproof against liquid molecules but harder to make the material vapour proof against the much smaller and more energetic molecules of a gas.

The permeability of materials to water vapour can be expressed in a number of ways.

Vapour resistivity

- *VAPOUR RESISTIVITY* (r_v) **is a measure of the resistance to the flow of water vapour offered by unit thickness of a particular material under standardised conditions**

UNIT: GN s/kg m

This property of a material is sometimes alternatively expressed as a *vapour diffusivity* or *vapour permeability* value, which is the reciprocal of vapour resistivity. Table 4.7 lists typical values for the vapour resistivity or vapour resistance of various building materials.

Table 4.7 *Vapour transfer properties of materials*

Material	Vapour resistivity (MN s/g m)
Brickwork	25–100
Concrete	30–100
Fibre insulating board	15–16
Hardboard	450–750
Mineral wool	5
Plaster	60
Plasterboard	45–60
Plastics	
expanded polystyrene	100–600
foamed polyurethane	30–1000
foamed urea formaldehyde	20–30
Plywood	1500–6000
Timber	45–75
Stone	150–450
Strawboard, compressed	45–75
Wood wool	15–40
Membrane	*Vapour resistance (GN s/kg)*
Aluminium foil	4000+
Bitumenised paper	11
Polythene sheet (0.06 mm)	125
Paint gloss (average)	6–20
Vinyl wallpaper (average)	6–10

Vapour resistance

- **VAPOUR RESISTANCE (R_v) describes the resistance of a specific thickness of material**

Vapour resistance is calculated by the following formula:

$$R_v = r_v L$$

where R_v = vapour resistance of that material (GN s/kg)
    L = thickness of the material (m)
    r_v = vapour resistivity of the material (GN s/kg m).

A vapour resistance value is usually quoted for thin membranes, such as aluminium foil or polythene sheet, and some typical figures are given in table 4.7.

Total vapour resistance (R_{vT}) of a compound structure is the sum of the vapour resistances of all the separate components:

$$R_{vT} = R_{v1} + R_{v2} + R_{v3} + \ldots \text{etc.}$$

Dew point gradients

The changes in temperature inside a structure such as a wall were calculated in chapter 2. The temperature change across any particular component is given by the formula:

$$\Delta\theta = \frac{R}{R_T} \times \theta_T$$

where $\Delta\theta$ = temperature difference across a particular layer
   R = resistance of that layer
   $\Delta\theta_T$ = total temperature difference across the structure
   R_T = total resistance of the structure.

The temperature drop across each component can be plotted onto a scale drawing of the structure to produce the temperature gradients, as shown in figure 4.7.

The *vapour pressure* drop across a component can be obtained in a similar manner from a similar formula:

$$\Delta P = \frac{R_v}{R_{vT}} \times P_T$$

where ΔP = vapour pressure drop across a particular layer
   R_v = vapour resistance of that layer
   P_T = total vapour pressure drop across the structure
   R_{vT} = total vapour resistance of the structure.

Vapour pressure changes can be plotted as gradients but they are usually converted to dew point readings and plotted as dew point gradients. The dew point at each boundary in the structure is obtained from a psychrometric chart by using the corresponding structural temperature and vapour pressure.

Worked example 4.4

An external wall is constructed with an inside lining of plasterboard 10 mm, then expanded polystyrene board (EPS) 25 mm, then dense concrete 150 mm. The thermal resistances of the components, in m² K/W, are: internal surface resistance 0.123, plasterboard 0.06, EPS 0.75, concrete 0.105, and external surface resistance 0.055. The vapour resistivities of the components, in MN s/g m, are: plasterboard 50, EPS 100, and concrete 30. The inside air is at 20 °C and 59 per cent RH; the outside air is at 0 °C and saturated. Use a scaled cross-section diagram of the wall to plot the structural temperature gradients and the dew point gradients.

Step 1: Use thermal resistances to calculate the temperature drops across each layer and the temperature at each boundary. Tabulate the information.

Layer	Thermal resistance (m²K/W)	Temperature drop $\left(\Delta\theta = \dfrac{R}{R_{\mathrm{T}}} \times \Delta\theta_{\mathrm{T}} \right)$	Boundary temperature (°C)
Inside air	–	–	20
Internal surface	0.123	$\dfrac{0.123}{1.093} \times 20 = 2.3$	–
Boundary	–	–	17.7
Plaster	0.06	$\dfrac{0.06}{1.093} \times 20 = 1.1$	–
Boundary	–	–	16.6
EPS	0.75	$\dfrac{0.75}{1.093} \times 20 = 13.7$	–
Boundary	–	–	2.9
Concrete	0.105	$\dfrac{0.105}{1.093} \times 20 = 1.9$	–
Boundary	–	–	1.0
External surface	0.055	$\dfrac{0.055}{1.093} \times 20 = 1.0$	–
Outside air	–	–	0.0
	$R_{\mathrm{T}} = \overline{1.093}$		

Total temperature drop $= 20 - 0 = 20\,°C$

Step 2: Plot the boundary temperatures on a scaled section of the wall and join the points to produce temperature gradients, as in figure 4.7 on page 111.

Step 3: Use vapour resistances to calculate the vapour pressure drops across each of the layers then, using the psychrometric chart, find the dew point temperature at each boundary (see table on page 111).

Inside vapour pressure $= 1400$
Outside vapour pressure $= 600$ from psychrometric chart

Total vapour pressure drop $= 1400 - 600$
$= 800\,Pa$

Step 4: Plot the dew point temperatures on the scaled, section diagram and produce dew point gradients, as in figure 4.7.

NOTE: Diagrams should be as large and as accurate as possible so they can be used to predict intermediate values of dew point at any place within the wall or roof.

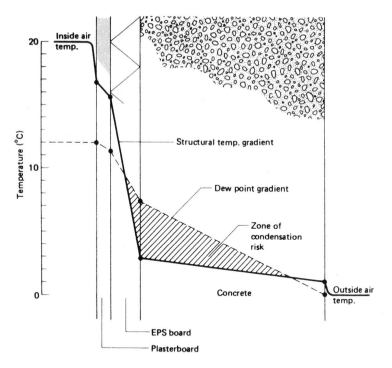

Figure 4.7 *Prediction of condensation*

Layer	Thickness L(m)	Vapour resistivity r_v	Vapour resistance $R_v = r_v L$	vp drop	vp at boundary (Pa)	Dew pt. at boundary
Internal				$\Delta P = \dfrac{R_v}{R_{vT}} \times P_T$		
surface	–	–	neg	–	–	–
Boundary	–	–	–	–	1400	12 °C
Plaster	0.010	50	0.5	$\dfrac{0.5}{7.5} \times 800 = 53$	–	–
Boundary	–	–	–	–	1347	11.5 °C
EPS	0.025	100	2.5	$\dfrac{2.5}{7.5} \times 800 = 267$	–	–
Boundary	–	–	–	–	1080	7.4 °C
Concrete	0.150	30	4.5	$\dfrac{4.5}{7.5} \times 800 = 480$	–	–
Boundary	–	–	–	–	600	0 °C
External						
surface	–	–	neg	–	–	–
			$R_{vT} = 7.5$			

111

Condensation risk

Condensation occurs on room surfaces, or within structures, if moist air meets an environment that is at or below the dew point temperature of that air. The prediction of such condensation risk is an important design technique for which assumptions are made about the air conditions expected inside and outside the building. The air temperatures and the dew points at any particular position within the structure can then be calculated and plotted.

Surface condensation
Alongside the surface of any partition in a building there is a layer of stationary air. The thermal resistance provided by this airlayer – the internal surface resistance – causes the temperature of the surface to be lower than the air temperature in the room. Consequently, air coming into contact with the surface will be cooled and may condense on the inside surface.

The surface temperature can be calculated using the following formula.

$$\Delta\theta = \frac{R}{R_T} \times \theta_T$$

In worked example 4.4, for example, the surface temperature of 17.7 °C is above the dew point of 12 °C, so surface condensation would not occur under these conditions.

Interstitial condensation
If moist air is able to permeate through a structure from inside to outside then it is usually cooled and follows the gradient of structural temperature change. The dew point temperature also falls and follows a gradient determined by the drop in vapour pressure across each material.

For specified air conditions, condensation will occur in any zone where the structural temperature gradient falls below the dew point gradient. These zones can be predicted by scaled diagrams, such as figure 4.7, where condensation will occur in the EPS layer and in the concrete.

Vapour barriers, vapour checks

The risk of interstitial condensation in structures is reduced if moist air is deterred from permeating through the materials of the structure.

- **A *VAPOUR BARRIER* or *VAPOUR CHECK* is a layer of building material which has a high resistance to the passage of water vapour**

It can be seen from table 4.7 of vapour resistance properties that no material is a perfect 'barrier' to the transfer of water vapour but some do offer

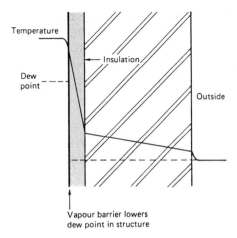

Figure 4.8 *Vapour barrier in wall*

an acceptably high resistance. Vapour barriers or checks can be broadly classified by the forms in which they are applied:

- **Liquid films:** examples include bituminous solutions, rubberised or siliconised paints, gloss paints.
- **Pre-formed membranes:** examples include aluminium-foil board, polythene-backed board, polythene sheet, bituminous felt, vinyl paper.

Vapour barriers need to be installed when there is a danger of interstitial condensation causing damage to the structure or to the insulation. Some types of interstitial condensation may be predicted and tolerated, such as when it occurs in the outer brick leaf of a cavity wall.

For a vapour barrier to be effective, it must block the passage of water vapour before the vapour meets an environment below the dew point temperature.

- Vapour barriers must be installed on the warm side of the insulation layer.

Poor installation can cause the real performance of a vapour barrier to be much worse than theory predicts. A small defect in one area of a vapour barrier degrades the performance of the entire vapour barrier, just like a 'small' leak in a gas mask. Incomplete seals at junctions, such as those between walls and ceilings, and punctures by pipes or electrical fittings are common defects.

Wall vapour barriers

Vapour barriers should be installed on the warm side of the insulation layer in a wall, as shown in figure 4.8. It is important that any water vapour inside the structure is able to escape and the outside surface should normally be permeable to water vapour, even though it may also need to be weatherproof. Various materials and constructions can be waterproof but still allow water vapour to pass because the molecules of liquid water are much larger than those of water vapour.

Roof vapour barriers

Roof cavities can easily suffer from severe condensation and appropriate ventilation and vapour barriers must be carefully provided.

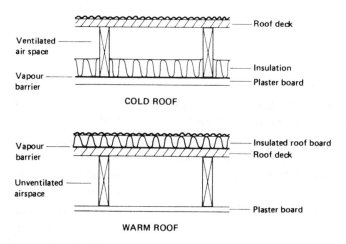

Figure 4.9 *Vapour barriers in roofs*

Cold roofs
If thermal insulation is installed at ceiling level then the remainder of the roof space will stay at a low temperature, which is a sign of effective insulation. However, this *cold roof* structure runs the risk of interstitial condensation because any water vapour passing through the roof will be cooled below its dew point temperature and condense into water liquid.

A vapour barrier must be installed on the warm side of the insulation in cold roof, as shown in figure 4.9. As a further precaution, the roof space must be ventilated to get rid of any vapour which does reach the roof space. In a flat roof, this ventilation must be carefully designed to achieve reliable flows of air and it may be easier to consider a warm roof design.

When the thermal insulation of a traditional pitched tile roof is upgraded by placing insulating material on top of the ceiling then the structure becomes a cold roof. There is usually enough accidental ventilation through the tiles and eaves to avoid condensation problems.

Warm roofs

A *warm roof* has its thermal insulation placed immediately beneath the waterproof covering and then protected with a vapour barrier, as shown in figure 4.9. The roof deck and the roof space therefore remain on the warm side of the insulation and will not become cool enough to suffer condensation. Any ceiling that is installed in a warm roof should have minimum thermal insulation so the temperature of the roof structure remains close to room temperature.

An *inverted roof* is a variation of warm roof design resulting from the installation of thermal insulating boards *above* the waterproof layer of the roof. Because all or part of the rainwater is designed to drain beneath the insulating boards this type of roof is also known as an 'upside-down roof'.

Exercises

1 A sample of air at 1 °C has a vapour pressure of 540 Pa. Calculate the RH of this air, given that the SVP of water is 1230 Pa at 10 °C.

2 The air temperature in a room is 20 °C and the dew point is 15 °C. Calculate the RH of this air, given that the SVP of water is 2340 Pa at 20 °C and 1700 Pa at 15 °C.

3 Use the psychrometric chart for the following problems.
   (a) If air has a dry bulb temperature of 20 °C and a wet bulb temperature of 15 °C then find the RH and the dew point.
   (b) If air at 16 °C dry bulb has an RH of 70 per cent then find the moisture content.
   (c) If air at 20 °C dry bulb has a vapour pressure of 7 mb then find the RH.

4 External air at 4 °C and 80 per cent RH is heated to 20 °C. Use the psychrometric chart to solve the following conditions.
   (a) The RH of the heated air.
   (b) The RH of the heated air if 0.006 kg/kg of moisture is added while heating.
   (c) The increase in vapour pressure between the two states.

5 The surface of a wall has a temperature of 11 °C when air at 14 °C first begins to condense upon the wall surface. Use the psychrometric chart to find the following.
   (a) The RH of the air
   (b) The reduction in moisture content necessary to lower the RH of the air to 50 per cent.

6 A 102 mm thick brick wall is insulated on the inside surface by the addition of 40 mm of mineral wool covered with 10 mm of plasterboard.

The thermal resistances, in $m^2 K/W$, are: external surface 0.055, brick-work 0.133, mineral wool 0.4, and internal surface 0.123. The vapour resistivities, in GN s/kg m, are: brickwork 60, mineral wool 5, and plasterboard 50. The inside air is at 20°C and 59 per cent RH; the outside air is at 0°C and 100 per cent RH.

(a) Calculate the boundary values of structural temperatures and dew points.

(b) Plot a structural temperature profile and a dew point profile on the same scaled cross-section diagram of the wall.

(c) Comment upon the above results.

Answers

1 44 per cent

2 72.6 per cent

3 (a) 58 per cent, 12°C; (b) 0.008 kg/kg; (c) 30 per cent

4 (a) 27 per cent; (b) 68 per cent; (c) 9.6 mb

5 (a) 81 per cent; (b) 0.003 kg/kg

5 Principles of Light

Light and the effects of light are a major element in the human sense of the environment. Both artificial and natural sources of light are used in buildings and these sources can be supplied and controlled in many ways. This chapter describes the nature of light, the effect of light upon the eye and the brain, and the measurement of light and lighting.

ELECTROMAGNETIC RADIATION

Light is energy in the form of electromagnetic radiation. This energy is radiated by processes in the atomic structure of different materials and causes a wide range of effects. The different forms of electromagnetic radiation all share the same properties of transmission although they behave quite differently when they interact with matter.

Light is that particular electromagnetic radiation which can be detected by the human sense of sight. The range of electromagnetic radiation to which the eye is sensitive is just a very narrow band in the total spectrum of electromagnetic emissions, as is indicated in figure 5.1.

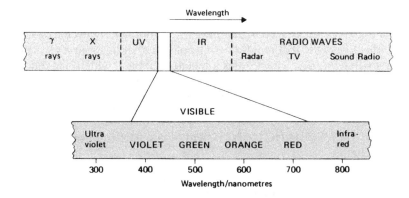

Figure 5.1 *Electromagnetic spectrum*

117

Electromagnetic waves

The transmission of light energy can be described as a wave motion or as packets of energy called photons. The two theories co-exist in modern physics and are used to explain different effects. The most convenient theory for everyday effects is that of electromagnetic wave motion. This can be considered as having the following general properties:

- The energy resides in fluctuations of electric and magnetic fields, which travel as a transverse wave motion
- The waves require no medium and can therefore travel through a vacuum
- Different types of electromagnetic radiation have different wavelengths or frequencies
- All electromagnetic waves have the same velocity, which is approximately 3×10^8 m/s in vacuum
- The waves travel in straight lines but can be affected by the following effects:

Reflection: Reversal of direction which occurs at a surface
Refraction: Deflection which occurs at the boundaries of different materials
Diffraction: Deflection which occurs at apertures or edges of objects.

Visible radiation

The wavelengths of electromagnetic radiation that are visible to the eye range from approximately 380 nm to 760 nm (1 nanometer (nm) is 10^{-9} metre). If all the wavelengths of light are seen at the same time the eye cannot distinguish the individual wavelengths and the brain has the sensation of white light.

- *WHITE LIGHT* **is the effect on sight of combining all the visible wavelengths of light**

White light can be separated into its component wavelengths. One method is to use the different refractions of light that occur in a glass prism, as shown in figure 5.2. The result is a spectrum of light, which is traditionally described in the seven colours of the rainbow although, in fact, there is a continuous range of hues (colours) whose different wavelengths cause different sensations in the brain.

- *MONOCHROMATIC LIGHT* **is light of one particular wavelength and colour**

If the colours of the spectrum are recombined then white light is again produced. Varying the proportions of the individual colours can produce different qualities of 'white' light.

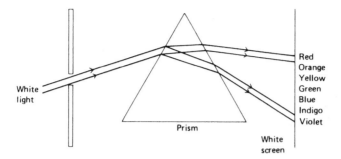

Figure 5.2 *Dispersion of white light*

Non-visible radiation

Electromagnetic radiations with wavelengths outside the range of visible wavelengths cannot, by definition, be detected by the human eye. However, those radiations immediately adjacent to the visible range of wavelengths are emitted by the Sun, along with light, and are often relevant to lighting processes.

Infra-red
Infra-red (IR) radiation has wavelengths slightly greater than those of red light and can be felt as heat radiation from the Sun and from other heated bodies. Infra-red radiation is made use of in radiant heating devices, for detecting patterns of heat emissions, for 'seeing' in the dark, and for communication links.

Ultra-violet
Ultra-violet (UV) radiation has wavelengths slightly less than those of violet light. It is emitted by the Sun and also by other objects at high temperature. Ultra-violet radiation helps keep the body healthy but excessive amounts can damage the skin and the eyes. The composition of the Earth's atmosphere normally protects the planet from excessive UV radiation emitted by the Sun.

Ultra-violet radiation can be used to kill harmful bacteria in kitchens and in hospitals. Certain chemicals can convert UV energy to visible light and the effect is made use of in fluorescent lamps.

NATURE OF VISION

The portion of the electromagnetic spectrum known as light is of environmental interest to human beings because it activates our sense of sight, or vision. Vision is a sensation caused in the brain when light reaches the eye. The eye initially treats light in an optical manner producing a physical image in the same way as a camera. This image is then interpreted by the brain in a manner which is psychological as well as physical.

The eye

Figure 5.3 shows the main features of the human eye with regard to its optical properties. The convex lens focuses the light from a scene to produce an inverted image of the scene on the retina. When in the relaxed position the lens is focused on distant objects. To bring closer objects into focus the cilary muscles increase the curvature of the lens; a process called *accommodation*. The closest distance at which objects can be focused, called the *near point*, tends to retreat with age as the lens becomes less elastic.

The amount of light entering the lens is controlled by the iris, a coloured ring of tissue, which automatically expands and contracts with the amount of light present. The retina, on which the image is focused, contains light receptors which are concentrated in a central area called the *fovea*; and which are deficient in another area called the *blind spot*.

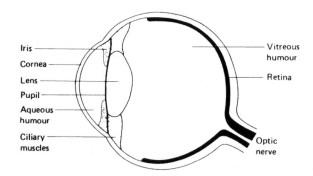

Figure 5.3 *Structure of the eye*

Operation of vision

The light energy falling on the retina causes chemical changes in the receptors which then send electrical signals to the brain via the optic nerve. A large portion of the brain is dedicated to the processing of the information received from the eyes and the eyes are useless if this sight centre in the brain is damaged.

The initial information interpreted by the brain includes the brightness and colour of the image. The stereoscopic effect of two eyes gives further information about the size and position of objects. The brain controls selection of the many items in the field of view and the sense of vision greatly depends on interpretations of images learned from previous experience.

Sensitivity of vision

The light-sensitive receptors on the retina are of two types. These receptors respond to different wavelengths of light in the manner shown in figure 5.4 and they give rise to two types of vision.

Cone vision
The cones are the light receptors that operate when the eye is adapted to normal levels of light. The spectrum appears coloured. There is a concentration of cones on the fovea at the centre of the retina and these are used for seeing details.

Rod vision
The rods are the light receptors that operate when the eye is adapted to very low levels of light. The rods are much more sensitive than the cones but the spectrum appears uncoloured. The colourless appearance of objects in moonlight or starlight is an example of this vision. There is a concentration of rods at the edges of the retina, which cause the eyes to be sensitive to movements at the boundary of the field of view.

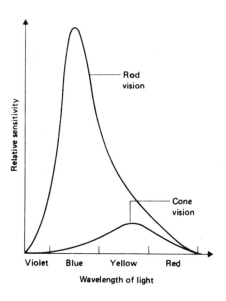

Figure 5.4 *Sensitivity of the eye*

Terminology of vision

Visual field
Visual field is the total extent in space that can be seen when looking in a given direction.

Visual acuity
Acuity is the ability to distinguish between details that are very close together. This ability increases as the amount of available light increases.

Adaption
Adaption is the process occurring as the eyes adjust to the relative brightness or colour of objects in the visual field. The cones and the rods on the retina take a significant amount of time to reach full sensitivity.

Contrast
Contrast is the difference in brightness or colours between two parts of the visual field.

MEASUREMENT OF LIGHTING

Light is one form of energy and could be measured by the standard units of energy. But the effect of light on the human environment also depends upon the sensitivity of the eye and special set of units has therefore been developed for the measurement of light and its effects.

Solid angle

As light can radiate in all three dimensions it is necessary to measure the way in which the space around a point can be divided into 'solid angles'. The standard SI unit of solid angle is the steradian, illustrated in figure 5.5.

- **One *STERADIAN* (ω) is that solid angle at the centre of a sphere which cuts an area on the surface of the sphere equal to the size of the radius squared**

The size of a solid angle does not depend upon the radius of the sphere or upon the shape of the solid angle. The total amount of solid angle contained around a point at the centre of a sphere is equal to the number of areas, each of size radius squared, which can fit onto the total surface area of a sphere. That is

$$\text{Total solid angle around a point} = \frac{\text{Surface area at sphere}}{\text{Area giving 1 steradian}} = \frac{4\pi r^2}{r^2}$$

$$= 4\pi$$

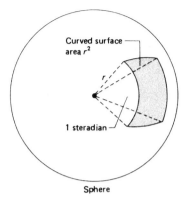

Figure 5.5 *The steradian*

Therefore a complete sphere contains a total of 4π steradians.

Luminous intensity

The concept of luminous intensity is used to compare different light sources and measure their 'strength'.

- **LUMINOUS INTENSITY (*I*) is the power of a light source, or illuminated surface, to emit light in a particular direction**

    UNIT: candela (cd)

The candela is one of the base units in the SI system. One candela is defined as the luminous intensity in a given direction of a source that emits monochromatic radiation of frequency 540×10^{12} Hz and of which the radiant intensity in that direction is $1/683$ W/ω.

The effect of one candela is still approximately the same as the original idea of one candlepower and the mean spherical intensity (MSI) of a 100 W light bulb, for example, is about 100 cd.

Luminous flux

The rate of flow of any electromagnetic energy can be expressed in terms of power but light energy is also measured by luminous flux.

- **LUMINOUS FLUX (*F*) is the rate of flow of light energy**

    UNIT: lumen (lm)

By definition, one lumen is the luminous flux emitted within one steradian by a point source of light of one candela, as shown in figure 5.6.

In general, luminous flux and luminous intensity are related by the following formula:

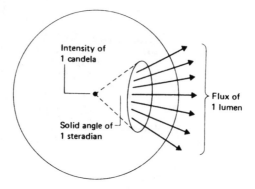

Figure 5.6 *Definition of the lumen*

$$I = \frac{F}{\omega}$$

where I = mean spherical intensity of the source (cd)
F = luminous flux emitted by the source (lm)
ω = solid angle containing the flux (sterad).

In the common case of point source emitting flux in all directions, the total solid angle around the point is 4π steradians.
Using

$$I = \frac{F}{\omega}$$

$$F = I \times \omega$$

gives the following useful formula:

$$F = I \times 4\pi$$

Illuminance

When luminous flux falls on a surface it illuminates that surface. The lighting effect is termed Illuminance.

- **ILLUMINANCE (E) is the density of luminous reaching a surface**

  UNIT: lux (lx) where 1 lux = 1 lumen/(metre)^2

Common luminance levels range from 50 lux for low domestic lighting to 50 000 lux for bright sunlight. Recommended lighting levels are specified in terms of illuminance and examples of standards are given in the section on lighting design.

If light is falling on a surface at right angles to the surface then the illuminance is given by the following formula:

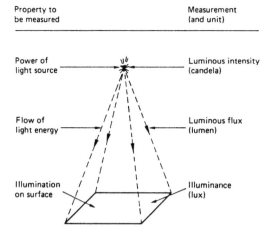

Property to be measured		Measurement (and unit)
Power of light source		Luminous intensity (candela)
Flow of light energy		Luminous flux (lumen)
Illumination on surface		Illuminance (lux)

Figure 5.7 *Summary of lighting measurements*

$$E = \frac{F}{A}$$

where E = illuminance on surface (lx)
      F = total flux reaching surface (lm)
      A = area of the surface (m²).

Worked example 5.1

A small source of light has a mean spherical intensity of 100 cd. One quarter of the total flux emitted from the source falls at right angles onto a surface measuring 3 m by 0.7 m. Calculate:

(a) the total luminous flux given out by the source; and
(b) the illuminance produced on the surface.

(a) Know I = 100 cd, $\omega = 4\pi$, F = ?

Using $I = \dfrac{F}{\omega}$

         $100 = \dfrac{F}{4\pi}$ or $F = 4\pi \times 100$

              $= 1256.64$

Total flux = **1256.64 lm**

(b) Know F = 1256.64 × 0.25 = 314.16 lm, A = 0.7 × 3 = 2.1 m², E = ?

Using $E = \dfrac{F}{A}$

$\qquad = \dfrac{314.16}{2.1} = 149.6$

So Illuminance = **150 lx**

Inverse square law of illumination

As the luminous flux emitted by a point source of light travels away from the source the area over which the flux can spread increases. Therefore, the luminous flux per unit area (i.e. the illuminance) must decrease. This relationship is expressed by the inverse square law, as illustrated in figure 5.8.

Inverse square law

- **The illuminance produced by a point source of light decreases in inverse proportion to the square of the distance from the source**

In SI units this law may be expressed mathematically by the following formula:

$$E = \frac{I}{d^2}$$

where I = intensity of a point source (cd)
$\qquad d$ = distance between source and surface (m)
$\qquad E$ = illuminance on that surface (lx).

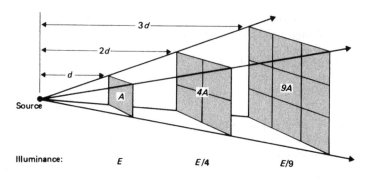

Figure 5.8 *Inverse square law of illumination*

An important consequence of the inverse square law is that changes in the position of light sources produce relatively large changes in lighting effect. For example, doubling the distance between a lamp and a surface causes the illuminance on that surface to decrease to one quarter of its original value.

Worked example 5.2

A lamp has a luminous intensity of 1200 cd and acts as a point source. Calculate the illuminance produced on surfaces at the following positions:

(a) At 2 m distance from the lamp, and
(b) at 6 m distance from the lamp.

Know: $I = 1200\,cd$, $d_1 = 2\,m$ and $d_2 = 6\,m$, $E = ?$

Using $E = \dfrac{I}{d^2}$

(a)

$$E = \frac{I}{d_1^2} \quad \text{or} \quad E = \frac{1200}{2^2} = 300$$

So illuminance at 2 m = **300 lx**

(b)

$$E = \frac{I}{d^2} \quad \text{or} \quad E = \frac{1200}{6^2} = 33.33$$

So illuminance at 6 m = **33.33 lx**

Cosine law of illumination

When the luminous flux from a point source reaches the surface of a surrounding sphere the direction of the light is always at right angles to that surface. However, light strikes many surfaces at an inclined angle and therefore illuminates larger areas than when it strikes at a right angle. The geometrical effect is shown in figure 5.9. If the luminous flux is kept constant but spread over a larger area then the illuminance at any point on that area must decrease.

Figure 5.10 shows how the area illuminated increases by a factor of 1/cosine θ. Because illuminance is equal to luminous flux divided by area the illuminance decreases by a factor of cosine θ. This relationship is sometimes termed *Lambert's Cosine Rule* and can be expressed by a general formula combining the factors affecting illumination:

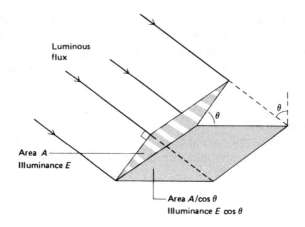

Figure 5.9 *Cosine law of illumination*

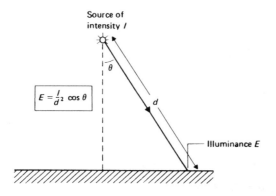

Figure 5.10 *Laws of illumination*

$$E = \frac{I}{d^2}\cos\theta$$

where E = illuminance on surface (lx)
I = intensity of source (cd)
d = distance between source and surface (m)
θ = angle between direction of flux and the normal.

Note: when $\theta = 0°$ then $\cos\theta = 1$.

The Cosine Law affects most practical lighting arrangements as it is usually difficult for all surfaces to receive light at right angles. For example, when a lamp on the ceiling illuminates a floor, only the point directly below the light fitting will receive luminous flux at right angles; the light will strike all other parts of the floor at varying angles of inclination.

The Cosine Law also has a large effect on *solar radiation* received by different locations and buildings. The radiation from the Sun strikes the surface of the Earth at different angles of incidence, which vary from the ideal right angle especially at higher latitudes away from the Equator. The intensity of light and heat received on flat ground decreases in accordance with the Cosine Law but can be increased by arranging surfaces at right angles to the incident radiation. The walls and windows of buildings therefore have higher solar gains than the roofs when the sun is low in the sky.

Worked example 5.3

A lamp acts as a point source with a mean spherical intensity of 1500 cd. It is fixed 2 m above the centre of a circular table which has a radius of 1.5 m. Calculate the illuminance provided at the edge of the table, ignoring reflected light.

Know $I = 1500\,cd$, $E = ?$

Using triangle laws

$$d^2 = 2^2 + 1.5^2$$
$$= 4 + 2.25 = 6.25$$
$$d = 2.50$$

$$\cos\theta = \frac{\text{opposite}}{\text{adjacent}} = \frac{2}{d} = \frac{2}{2.5}$$

Note: you do not need to know the actual value of the angle in degrees.

Using

$$E = \frac{I}{d^2}\cos\theta$$

$$E = \frac{1500}{6.25} \times 0.8 = 192$$

So illuminance at table edge = **192 lx**

Reflection

One method of changing the direction of light is by the process of reflection, which may be of two types as shown in figure 5.11:

- **Specular reflection** is direct reflection in one direction only. The angle of incidence (*i*) equals the angle of reflection (*r*).
- **Diffuse reflection** is reflection in which the light is scattered in various directions.

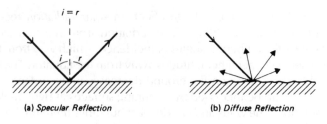

Figure 5.11 *Reflection of light*

Reflectance

Most practical surfaces give a mixture of free and diffuse reflection properties. The amount of reflection at a surface is measured by the *reflection factor* or reflectance.

- *REFLECTANCE* **is the ratio of the luminous flux reflected from a surface to the flux incident upon the surface**

  Typical values of reflectance are given in table 5.I.

- Maximum reflectance has a value near 1; for light, shiny surfaces
- Minimum reflectance has a value near 0; for dark, dull surfaces.

Table 5.1 *Reflectances of building surfaces*

Surface reflectance
White emulsion paint on plaster 0.8
White emulsion paint on concrete 0.6
Concrete: light grey 0.4
Timber: birch, beech, or similar 0.3
Bricks: red fletton 0.35
Quarry tiles: red 0.1

Reflected components of light are an important factor in illumination. The following is a simple worked example and the influence of reflectances will be referred to again in the section on lighting design.

Worked example 5.4

A point source of light with a luminous intensity of 1800 cd is set 3 m above the floor and 1 m below the ceiling which has a reflectance of 0.5. Calculate the direct and reflected components of the illuminance on the floor beneath the light.

(a) Know $I = 1800$ cd, $d = 3$ m, $E = ?$

Using

$$E = \frac{I}{d^2}$$

$$= \frac{1800}{3^2} = 200$$

So direct component = **200 lx**

(b) Know $I = 1800\,cd$, $d = 1 + 1 + 3 = 5\,m$, $E = ?$

Using

$$E = \frac{I}{d^2} \times \text{reflectance}$$

$$E = \frac{1800}{5^2} \times 0.5 = 36$$

So reflected component = **36 lx**

Luminance

The appearance of an object is affected by the amount of light it emits or reflects compared with the area of its surface. This idea of surface 'brightness' is termed luminance.

- *LUMINANCE (L)* **is a measure of the ability of an area of light source, or reflecting surface, to produce the sensation of brightness**

For example, a spot lamp and a fluorescent tube may each have the same luminous intensity. But, because of its larger surface area, the fluorescent tube would have a lower luminance. There are two types of Luminance measurement.

Self-luminous sources
Self-luminous sources include light sources and also reflecting surfaces, like the Moon, which can be considered as a source of light. The luminance is given by the following formula:

$$L = \frac{\text{Luminous intensity in a given direction}}{\text{Area of source as seen from that direction}}$$

UNIT: candela per square metre (cd/m^2)

Reflecting surfaces
For reflecting surfaces only, the luminance can be expressed in terms of the luminous flux emitted per unit area:

$$L = \frac{\text{Luminous flux reflected}}{\text{Area of reflecting surface}}$$

UNIT: apostilb (asb) where 1 asb = 1 lumen/m

The candela per square metre is the correct SI unit of luminance. The apostilb is an alternative unit of luminance which is convenient for some calculations. 1 apostilb = $1/\pi$cd/m^2. This luminance of a surface can be calculated as the product of the illuminance and the reflectance of a surface.

Glare

The eye can detect a wide range of light levels but vision is affected by the range of brightness visible at any one time.

- *GLARE* **is the discomfort or impairment of vision caused by an excessive range of brightness in the visual field**

Glare can be caused by lamps, windows, and painted surfaces appearing too bright in comparison with the general background. Glare can be further described as disability and discomfort glare.

Disability glare
Disability glare is the glare that lessens the ability to see detail. It does not necessarily cause visual discomfort. For example, excessive reflections from shiny white paper can cause disability glare while reading.

Discomfort glare
Discomfort glare is the glare that causes visual discomfort without necessarily lessening the ability to see detail. An unshielded light bulb is a common example of discomfort glare. The amount of discomfort depends on the angle of view and the type of location. If the direction of view is fixed on a particular visual task, then glare caused by lighting conditions will be more noticeable.

Light meters

A *photometer* is an instrument that measures the luminous intensity of a light source by comparing it with a standard source whose intensity is known. The distances between the instrument and the two light sources are adjusted until they each provide the same illuminance at the photometer.

The human eye is the best judge of this equal illuminance and photometers generally use some system which allows two screens to be compared. The inverse square law can be used to calculate the unknown intensity. Under conditions of equal illuminance the following formula is true:

$$\frac{I_1}{d_1^2} = \frac{I_2}{d_2^2}$$

where I_1 = intensity of source at distance d_1
      I_2 = intensity of source at distance d_2.

A *photocell* light meter is an instrument that directly measures the illuminance on a surface. The electrical resistance of some semiconductors, such as selenium, changes with exposure to light and this property is used in an electrical circuit connected to a galvanometer. This meter may be calibrated in lux.

Directional quality

The appearance of an object is affected by the direction of light as well as by the quantity of light illuminating the object.

- **Illumination vector:** the quantity of light from a specified direction, such as horizontal illumination.
- **Scalar illuminance:** the total illuminance caused by light from all directions, including reflected light.
- **Vector/scalar ratio:** a measure of the directional strength of light at a particular point.

A vector/scalar ratio of 3.0 indicates lighting with very strong directional qualities, such as that provided by spotlights or direct sunlight. The effect is of strong contrasts and dark shadows. A vector/scalar ratio of 0.5 indicates lighting with weak directional qualities, such as that provided by indirect lighting by reflections – this lighting is free of shadows and makes objects look 'flat'.

COLOUR

Colour is a subjective effect that occurs in the brain when the eye is stimulated by various wavelengths of light. It is difficult to specify colour by any method and especially difficult by means of this black and white printing. But the description and measurement of colour are important in the design of lighting schemes, as well as in photographic films, paints, dyes, and inks.

Spectral energy distribution

The brain can experience the impression of only one colour at a time and it cannot detect how that colour has been made up. It is possible for different combinations of different wavelengths to produce the same colour impression in the brain.

A *spectrometer* is an instrument that disperses light into its component wavelengths and measures the amount of light energy radiated at each wavelength. This spectral energy distribution reveals if some light is being emitted at each wavelength (continuous spectrum), or only at certain wavelengths (discontinuous or line spectrum). Some typical spectral curves are shown in figure 5.12.

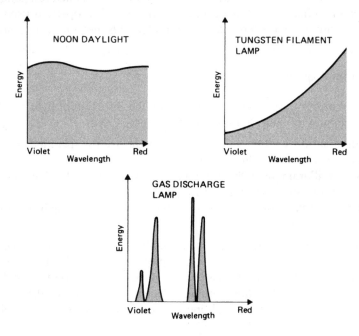

Figure 5.12 *Spectral energy distribution*

Colour systems

CIE colour coordinates
The Commission Internationale de l'Eclairage (CIE) has produced many international lighting standards. The CIE coordinate system describes any colour as a mixture of three monochromatic primary colours. Three coordinates specify how much of each primary colour needs to be mixed to reproduce the colour being described.

Munsell system
The Munsell system is used by some architects and interior designers for classifying the colour of surfaces. It specifies colours in terms of three factors called hue (basic colour), chroma (intensity of colour), and value (greyness).

Colour reproduction

The human vision can distinguish many hundreds of different colours and intensities of colour. Modern systems of colour representation by ink, photography, and television can reproduce these many colours to the satisfaction of the eye, yet such systems use just a few basic colours.

It has been shown that white light contains all the colours of the spectrum which can be recombined to give white light again. White light can also be split into just three colours which recombine to give white light. In addition to white light, *any* colour can be reproduced by various combinations of three suitable primary colours. Most colour systems use this *trichromatic* method of reproduction and the eye is believed to send its information to the brain by a similar method of coding.

Newton's discoveries concerning the combination of colours initially seemed to disagree with the experience of mixing paints. This confusion about combining colours can still arise because they can be mixed by two different methods which have different effects: additive mixing and subtractive mixing.

Additive colour

If coloured lights are added together then they will produce other colours. When the three primary additive colours are combined in equal proportions they *add* to produce white light.

Additive primary colours
- Red + Green + Blue = White
- Red + Green   = Yellow
-        Green + Blue = Cyan
- Red +   Blue = Magenta

Note: Cyan is a sky-blue colour; Magenta is a purple-red colour.

Applications of additive colour
Some of the important applications of additive colour mixing are described below:
- **Stage lighting:** By using three or more coloured light sources, on dimming controls, any colour effect can be obtained.
- **Colour television:** A video screen has many small red, green, and blue phosphors, each type controlled by a separate beam of electrons. A close look at a colour television will reveal that white doesn't actually exist on the screen but is an effect produced in the brain.
- **Colour printing:** Some processes, such as gravure, produce a mosaic of ink dots which effectively act as an additive system.

Subtractive colour

If colours are subtracted from white light then other colours will be produced. When the three primary subtractive colours are combined in equal proportions they *subtract* (absorb) components of white light to produce darkness.

Subtractive primary colours

Cyan subtracts Red
Magenta subtracts Green   These pairs are termed *complementary*
Yellow subtracts Blue   colours

White light can be considered as a combination of red, green, and blue light. Materials that transmit or reflect light absorb selected wavelengths and pass the remaining light to the eye. Combining two or more paints, or coloured layers, has a cumulative effect, as shown in figure 5.13.

A red surface is defined by the fact that it subtracts green and blue from white light, leaving only red light to reach the eye. A white surface reflects all colours. If the colour content of the light source changes then the appearance of the surface may change. This effect is important when comparing surface colours under different types of light source.

Applications of subtractive colour

Most colour is seen as a result of subtractive processes and some common applications are given below:

- **Paint pigments:** Paint colours are mixed according to subtractive principles, even though, for simplicity, the primary colours are often called red, yellow, and blue instead of magenta, yellow, and cyan.
- **Colour photographs:** All colour transparencies and colour prints are finally composed of different densities of three basic dyes: cyan, magenta, and yellow.

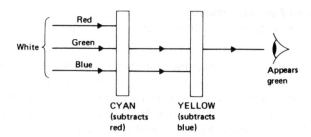

Figure 5.13 *Combination of subtractive colours*

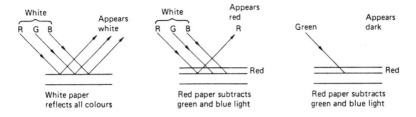

Figure 5.14 *Subtractive colour appearance*

- **Colour printing:** White paper is over-printed three times with the three basic colours of printing ink which are cyan, magenta, and yellow. Black ink is also used to achieve extra density.

Exercises

1 A small lamp emits a total luminous flux of 1257 lm in all directions. Calculate the luminous intensity of this light source.

2 A point source of light has an intensity of 410 cd and radiates uniformly in all directions.
   (a) Calculate the quantity of flux flowing into a hemisphere.
   (b) Calculate the average illuminance produced on the inside surface of this hemisphere if it has a radius of 1.5 m.

3 A small lamp has a mean luminous intensity of 80 cd. Calculate the maximum direct illuminance the lamp produces on a surface under the following conditions:
   (a) At a distance of 0.8 m from the lamp.
   (b) At a distance of 3.2 m from the lamp.

4 A street lamp has a uniform intensity of 1200 cd. It is positioned 7 m above the centre line of a road which is 8 m wide.
   (a) Calculate the illuminance on the road surface directly below the lamp.
   (b) Calculate the illuminance at the edge of the roadway.

5 A uniform point source of light emits a total flux of 2500 lm. It is suspended 800 mm above the centre of a square table with sides of length 600 mm. Calculate the minimum and maximum illuminances produced on the table.

6 A photometer is positioned on a direct line between two lamps. When each inside of the photometer receives equal illuminance the distances

from the photometer are 500 mm to lamp A, and 650 mm to lamp B. Lamp A is known to have a luminous intensity of 70 cd. Calculate the luminous intensity of lamp B.

7 Use a diagram to predict and demonstrate the appearance of white light after it has passed through a cyan glass and then a magenta glass. Assume that white light is simply composed of red, green, and blue.

8 Use a diagram to predict and demonstrate the appearance of a blue surface which is illuminated by yellow light.

Answers

1 100 cd

2 (a) 2576 lm; **(b)** 182 lx

3 (a) 125 lx; **(b)** 7.81 lx

4 (a) 24.5 lx; **(b)** 16.03 lx

5 310.9 lx, 214.3 lx

6 118.3 cd

6 Artificial Lighting

The type of lighting chosen for a building is closely linked to other design decisions for the building, such as the basic plan shape, the type and extent of windows, and the type of heating or cooling. The subject of this chapter is artificial lighting, and the following chapter deals with natural lighting. Although the principles of these two topics are treated separately it is important that they are considered together when designing a building.

The main functions of artificial lighting can be summarised as follows:

- **Task:** To provide enough light for people to carry out a particular activity.
- **Movement:** To provide enough light for people to move about with ease and safety.
- **Display:** To display the features of the building in a manner suitable for its character and purpose.

To achieve these aims it is necessary to consider the properties of lamps, of the lamp fittings, and of the room surfaces that surround them.

LAMPS

The oldest source of artificial light is the flame from fires, from candles, and from oil lamps where light is produced as one of the products of chemical combustion. Modern sources of artificial light convert electrical energy to light energy and are of two general types: incandescent sources and gas discharge sources.

Incandescent lamps
Incandescent sources produce light by heating substances to a temperature at which they glow and are luminous. Incandescence can be achieved by heating with a flame but in an electric lamp, such as the light bulb, a metal wire is heated by an electric current.

Discharge lamps
Discharge lamps produce light by passing an electric current through a gas or vapour that has become ionised and hence able to conduct electricity. At

low gas pressures, a luminous arc or discharge is formed between the electrodes and useful quantities of light are given off. Discharge lamps need special control gear in their circuits and the colour quality of their light is often poor.

The characteristics of the electric lamps used in modern lighting are summarised in table 6.1. Further details of these lamps and their properties are given m the following sections.

Properties of lamps

Luminous efficacy
The ability of a lamp to convert electrical energy to light energy is measured by its efficacy which is given by the following formula:

$$\text{Luminous efficacy} = \frac{\text{Luminous flux output}}{\text{Electrical power input}}$$

UNIT: lumens/watt (lm/W)

The electrical running costs of a lamp can be calculated from its efficacy. The luminous efficacy of a lamp varies with its type and its wattage so exact data should be obtained from the manufacturer.

Life
The luminous efficacy of a lamp decreases with time and for a discharge lamp it may fall by as much as 50 per cent before the lamp fails. The nominal life of a lamp is usually determined by the manufacturer by considering the failure rate of a particular model of lamp combined with its fall in light output. In a large installation it is economically desirable that all the lamps are replaced at the same time on a specified maintenance schedule.

Colour temperature
The qualities of light emitted by heated objects depend upon the temperature of the radiating object and this fact can be used to describe the colour of light. A theoretically perfect radiator, called a 'black body', is used as the standard for comparison.

- **The *CORRELATED COLOUR TEMPERATURE (CCT)* of a light source is the absolute temperature of a perfect radiator when the colour appearance of the radiator best matches that of the light source**

    UNIT: Kelvin (K)

This method of specifying colour quality is most suitable for light sources that emit a continuous spectrum, such as those giving various types of 'white' light. The lower values of colour temperature indicate light with a higher red content. Some examples of colour temperatures are given below:

Clear sky	12000–24000 K
Overcast sky	5000–8000 K
Tubular fluorescent lamps	3000–6500 K
Tungsten filament lamps	2700–3100 K

Colour rendering

The colour appearance of a surface is affected by the quality of light from the source. Colour rendering is the ability of a light source to reveal the colour appearance of surfaces. This ability is measured by comparing the appearance of objects under the light source with their appearance under a reference source such as daylight.

One system specifies the colour rendering of lamps by a *colour rendering index* (R_a) which has a value of 100 for an ideal lamp. Practical sources of white light range in R_a value between 50 and 90.

Table 6.1 *Characteristics of electric lamps*

Lamp type (code)	Wattage range	Typical efficacy (lm/W)	Nominal life (hours)	Colour temperature (K)	Typical applications
Tungsten filament (GLS)	40–200	12	1000	2700	Homes, hotels and restaurants
Tungsten–halogen (T-H)	300–2000	21	2000–4000	2800–3000	Area and display lighting
Compact fluorescent (CFL)	9–20	60	8000	3000	Homes, offices and public buildings
Tubular fluorescent (MCF)	20–125	60	8000+	3000–6500	Offices and shops
Mercury fluorescent (MCB)	50–2000	60	8000+	4000	Factories and roadways
Mercury halide (MBI)	250–3500	70	8000+	4200	Factories and shops
Low-pressure sodium (SOX)	35–180	180 (at 180 W)	8000+	n a	Roadways and area lighting
High-pressure sodium (SOX)	70–1000	125 (at 400 W)	8000+	2100	Factories and roadways

Note
Higher wattage lamps generally have higher efficiency and longer life.
Values for the efficacy and life of most discharge lamps are subject to improvements – see manufacturers' data.

Tungsten filament lamps

Electric incandescent lamps work by passing an electric current through a filament of metal and raising the temperature to white heat. When the metal is incandescent, at around 2800 K, useful quantities of light are given off. Tungsten is usually used because it has a high melting point and low rate of evaporation.

To prevent oxidation (burning) of the metal the tungsten coil is sealed inside a glass envelope and surrounded by an inert atmosphere of unreactive gases such as argon and nitrogen. During the operation of the lamp tungsten is evaporated from the filament and deposited on the glass causing it to blacken. The filament therefore thins and weakens and must eventually break.

The simple tungsten lamp, such as a light bulb, is the oldest, shortest-lived, and least efficient type of electrical light source and is being replaced by more efficient lamps. But the properties of filament lamps have been greatly improved by using halogen gases and lower voltages, as described below, and these lamps are useful in modern lighting design.

General lighting service lamp

The general lighting service (GLS) lamp, or common light bulb, has a coiled filament contained within an envelope (bulb) of glass which may be clear or frosted. Inside the lamp is a fuse and typical construction is shown in figure 6.1.

The filament lamp produces a spectral distribution of light which is continuous but deficient in blue, as shown previously in figure 5.11. This

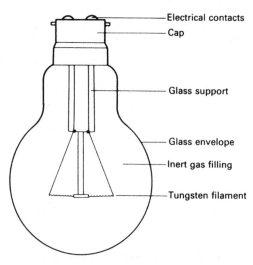

Electrical contacts
Cap
Glass support
Glass envelope
Inert gas filling
Tungsten filament

Figure 6.1 *Tungsten filament lamp*

quality of light is seen as 'warm' and is considered generally suitable for social and domestic applications.

The cost of a tungsten filament lamp is low and its installation is simple, but the relatively short life of the lamp can cause the labour costs of replacement to be high. We may not notice the labour of changing a 'light bulb' at home but it can be significant in large buildings and for high installations such as on a high ceiling. The over-riding cost factor against the tungsten filament is that the low luminous efficacy produces high electrical running costs. Only about 5 per cent of the electrical energy is converted to visible light and most of the energy consumed is given off as heat, especially radiant (infra-red) heat.

Reflecting lamps

The relatively large size of standard tungsten filament lamp makes it difficult to control the direction of the light. *Spotlamps* (PAR) are filament lamps with the glass bulb silvered inside and shaped to form a parabola with the filament at the focus. This arrangement gives a directional beam of light which is available in different widths of beam. *Sealed beam lamps* use similar techniques. *Crown-silvered lamps* (CSL) are standard filament lamps where the glass bulb is silvered in front. When this lamp is used with a special external reflector it also gives narrow beams of light.

Tungsten–halogen lamps

Tungsten–halogen lamps have filaments which run at higher temperatures with the presence of a small quantity of a halogen gas, such as iodine or bromine. When tungsten evaporates from the filament it is deposited on the hot wall of the lamp where it combines with the iodine. This new compound is a vapour which carries the tungsten back onto the lamp and re-deposits it on the hot filament, while the iodine is also re-cycled.

In order to run at higher temperatures the envelop of tungsten–halogen lamp is made of *quartz* instead of plain glass. The heat-resistance of the quartz allows the construction of a very small bulb for applications such as spotlamps, projectors, and car headlamps where directional control of light is important. Tungsten–halogen lamps also have the general advantages over simple tungsten lamps of increased efficiency and longer life.

Low-voltage systems

Tungsten halogen techniques have allowed the development of low-voltage bulbs where, because a lower resistance is needed, the filaments can be shorter, thicker, and stronger. A common system uses 12 volt lamps fed from the mains by a transformer.

The small size of these lamps gives them good directional qualities which make them popular in shops for the display of goods. The relatively low heat output of low-voltage systems is also an important property in stores where high levels of illumination can cause overheating.

Fluorescent lamps

Fluorescent gas discharge lamps work by passing an electric current through a gas or vapour so that a luminous arc is established within a glass container. The energised gas atoms emit ultra-violet (UV) radiation and some blue-green light. A coating of fluorescent powders on the inside of the glass absorbs the UV radiation and re-radiates this energy in the visible part of the spectrum. The fluorescent coating therefore increases the efficiency of the system and allows the colour quality of the light to be controlled.

Tubular fluorescent lamps

The common tubular fluorescent (MCF) lamp is a form of gas discharge lamp using mercury vapour at low pressure. Figure 6.2 illustrates the construction of a typical lamp and an example of the electrical control gear that it requires. This gear is needed to provide a starting pulse of high voltage, to control the discharge current, and to improve the electrical power factor. Modern types of control gear use electronic circuits instead of wire-wound components.

The colour quality of the light from a fluorescent lamp can be varied by using suitable mixtures of the metallic phosphors which make up the fluorescent coating. Lamps are available with colour temperatures ranging from 3000 K ('warm') to 6500 K ('daylight'). The large surface area of this type of

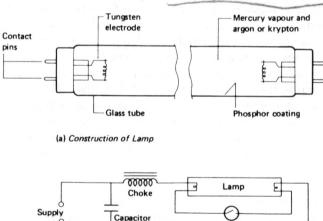

(a) *Construction of Lamp*

(b) *Typical Control Circuit*

Figure 6.2 *Tubular fluorescent lamp*

lamp produces lighting of a relatively non-directional 'flat' quality and with low glare characteristics.

Like all discharge lamps, fluorescent lamps continuously decrease in light output and efficacy as they are run and the lamp should be replaced after a stated number of hours. The lamp will usually run for longer than its stated life but the light output will then fall below the levels specified in the lighting design. The exact life of a discharge lamp also depends on how often the lamp is switched on or off.

It is possible for the cyclic nature of the gas discharge to be annoying and to cause a stroboscopic effect – an apparent change of motion, when viewing moving objects such as machinery. These effects are avoided by the use of modern electronic control gear operating at high frequency and the use of new lamps with shielded electrodes and high-efficiency phosphors.

The luminous efficacy of a tubular fluorescent lamp is at least five times better than that of a tungsten filament lamp, and modern types of narrow-diameter tube have even higher efficacies. The initial cost of a fluorescent lamp is higher than that of the tungsten filament lamp but this is soon offset by the lower running costs arising from their longer life, lower labour costs for changing lamps, and cheaper electricity costs.

Compact fluorescent lamps

Fluorescent lamps are available in compact forms comparable in size to a traditional tungsten filament lamp. Some makes have the electronic control gear incorporated inside the lamp so that they can be installed in a conventional light fitting to directly replace a tungsten filament lamp; other makes have the control gear in the fitting. Compact fluorescent lamps have a quality suitable for domestic and offices purposes and the use of such lamps are an important technique for low-energy lighting in buildings.

The current cost of a compact fluorescent lamp is around 10 times the cost of the light bulb that it will replace. Although householders are often reluctant to make the investment, a calculation such as that in worked example 6.1 shows how to calculate the saving of electrical energy and money over several years. Table 6.2 shows how a compact fluorescent lamp which delivers the same amount of light as a tungsten filament light bulb only uses about one fifth the amount of electrical energy or money.

Table 6.2 *Energy rating of comparable lamps*

Tungsten filament light bulbs	40 W	60 W	75 W	100 W
Equivalent compact fluorescent lamps	9 W	11 W	15 W	20 W

Note
The lamps in each vertical group give approximately the same light output in lumens.
The energy used and cost are directly proportional to the power rating in watts (W).

Worked example 6.1

A certain space needs to be illuminated for 20 hours a day with a total of 18 000 lm. Two types of lamp, detailed below, are considered in the design using a costing period of two years. For the two possible lamp systems:
(a) calculate the intial costs and running costs,
(b) compare the total costs of the two systems over 2 years.

Given data

	LAMP A	LAMP B
Lamp type:	tungsten filament	tubular fluorescent
Lamp wattage:	100 W	60 W
Lamp efficacy:	12 lm/W	60 lm/W
Lamp life:	1000 hours	8000 hours
Installation costs: (luminaire and gear)	£2 each lamp	£10 each lamp
Lamp costs: (parts and labour)	£0.70 each	£3 each
Electricity costs: (same each lamp)	8 pence = £0.08 per kilowatthour	

Working

	LAMP A	LAMP B
Total output per per lamp:	100 × 12 = 1200 lumens	60 × 60 = 3600 lumens
so Number of installations needed:	$\frac{18\,000}{1200} = 15$ units	$\frac{18\,000}{3600} = 5$ units
so Installation costs:	£2 × 15 = **£30**	£10 × 5 = **£50**
Total hours: (same each lamp)	2 yrs × 365 days × 20 hrs = 14 600 hrs	
Times each lamp replaced:	$\frac{14\,600}{1000} = 14.6$ or 15	$\frac{14\,600}{8000} = 1.83$ or 2
so Replacement costs:	£0.70 × 15 × 15 = £157.50	£3 × 5 × 2 = £30
Electrical energy for each lamp:	100/1000 kW × 14 600 h = 1460 kWh	60/1000 kW × 14 600 h = 876 kWh
Total Electricity:	1460 × 15 = 21 900 kWh	876 × 5 = 4380 kWh
so Electricity costs:	£0.08 × 21 900 = £1752.50	£0.08 × 4380 = £350.40

Running costs: (replacement + electricity)	£157.50 + £1752 = **£1909.50**	£45 + £350.40 = **£395.40**
Total costs over 2 years:	£30 + £1909.50 = **£1939.50** for tungsten filament lamps	£50 + £395.40 = **£445.40** for tubular fluorescent lamps

Discharge lamps

Apart from the well-known tubular fluorescent lamp, gas discharge lamps have in the past been restricted to outdoor lighting, such as for roadways, where their generally poor colour qualities have not been important. Modern types of discharge lamp have a colour rendering that is good enough for large-scale lighting inside buildings such as factories and warehouses. Continuing technical advances are producing more discharge lamps suitable for interior lighting and the high efficacy of such lamps can give significant savings in the energy use of buildings.

Mercury discharge lamps
An uncorrected mercury lamp emits sharp spectral peaks of light at certain blue and green wavelengths. A better spectral distribution is obtained by coating the glass envelope with fluorescent powders (MBF lamps).

In the *mercury halide* (MBI) lamp, metallic halides are added to the basic gas discharge in order to produce better colour rendering and to raise the efficacy.

Sodium discharge lamps
Low-pressure sodium (SOX) lamps produce a distinctive yellow light that is virtually monochromatic and gives poor colour rendering. However, the efficacy of the lamps is very high and they have been traditionally used for street lighting.

High-pressure sodium (SON) lamps produce a continuous spectrum without much blue light but with a colour rendering that is more acceptable than the low-pressure sodium lamp. SON lamps are used in modern street lighting and for the economic lighting of large areas such as forecourts and warehouses.

Lamp developments

Lighting is an important area of energy conservation in buildings as new systems can be installed in existing buildings as well as in new buildings.

The energy saved by a modernised lighting usually pays for the cost of the installation within a few years, which is a shorter payback period than most methods of saving energy in buildings. The major lamp companies of the world are therefore developing new types of low-energy lamps including improvements to the compact fluorescent lamp and high pressure discharge lamps described above.

Induction lamps are a completely new type of electric lamp which have neither filament nor electrodes. An electric induction coil operating at high frequency induces an energy flow which excites a gas, such as mercury, to give off ultra-violet energy. Visible light is then generated when the photons strike a fluorescent coating, as in existing gas discharge lamps. The advantages of the induction lamps are a high quality of white light and exceptionally long life, such as 60 000 hours.

LUMINAIRES

A luminaire is the light fitting that holds or contains a lamp. Luminaires usually absorb and redirect some of the luminous flux emitted by the lamp and, in the design of lighting installations, the choice of lamp must be combined with the choice of luminaire. The specification of luminaires is therefore important and this section outlines the methods used to classify luminaires according to the effect that they have on light distribution.

Luminaires may also serve a number of mechanical and electrical purposes such as positioning the lamp in space, protecting the lamp, and containing the control gear. Physical properties that may be relevant in the choice of the luminaire include its electrical insulation, moisture resistance, appearance, and durability.

Polar curves

Polar curves show the directional qualities of light from a lamp or luminaire by a graphical plot onto polar coordinate paper, as shown in figure 6.3. The luminous intensity of a lamp in any direction can be measured by means of a photometer, the results plotted and joined by a curve which then represents the distribution of light output from the fitting. If the distribution is not symmetrical about the vertical axis, as in a linear fitting, then more than one vertical plane needs to be plotted.

Light output ratio

It is convenient to try and describe the distribution of light from a luminaire by a system of numbers. One system is to classify luminaires by the propor-

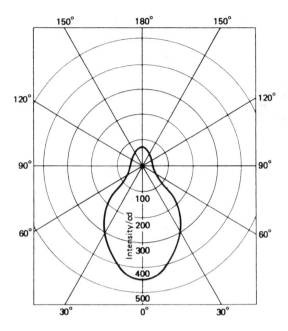

Figure 6.3 *Polar curve of a luminaire*

tion of the total light from the luminaire emitted into the upper and lower hemispheres formed by a plane through the middle of the lamp filament. These classifications are as follows.

Light output ratio

$$LOR = \frac{\text{Total light output of luminaire}}{\text{Light output of its lamp(s)}}$$

Downward light output ratio

$$DLOR = \frac{\text{Downwards light output of luminaire}}{\text{Light output of its lamp(s)}}$$

Upward light output ratio

$$ULOR = \frac{\text{Upwards light output of luminaire}}{\text{Light output of its lamp(s)}}$$

So that

$$LOR = DLOR + ULOR$$

Luminaires can be broadly divided into five types according to the proportion of light emitted upwards or downwards. Figure 6.4 indicates these divisions using luminaires with generalised shapes.

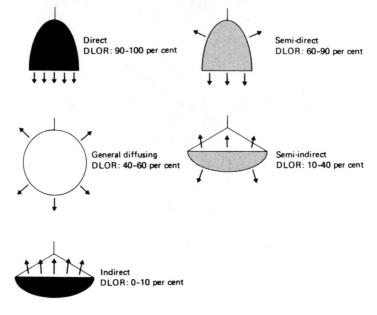

Figure 6.4   *General types of luminaire*

British zonal system

The British Zonal or BZ System classifies luminaires into 10 standard light distributions with BZ numbers from 1 to 10. BZ1 indicates a distribution of light that is mainly downwards and BZ10 a distribution that is mainly upwards. The BZ number is obtained from a combination of the properties of the luminaire and the room surfaces which are measured by the following terms:

- **Direct ratio** is the proportion of the total downward flux from the luminaires which falls directly on the working plane
- **Room index** is a number which takes account of the proportions of the room by the following formula

$$RI = \frac{L \times W}{H_m(L+W)}$$

where L = length of the room
W = width of the room
H_m = mounted height of the luminaire above the working plane.

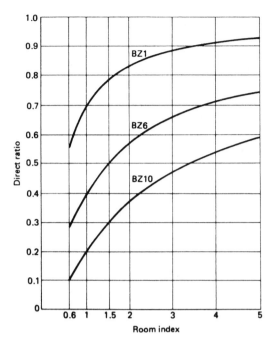

Figure 6.5 *BZ curves for luminaires*

A plot of direct ratio against room index for a particular luminaire gives a curve which can be described by the BZ number of the standard curve to which it is the closest fit. Figure 6.5 illustrates the standard curves of some BZ numbers. The actual curves of most luminaires cross from one BZ zone to another as the room index varies.

LIGHTING DESIGN

Illuminance levels

The quantity of light on a certain surface is usually the primary consideration in the design of a lighting system. This quantity is specified by the density of luminous flux, or illuminance, and measured in lux. The illuminance level changes with time and varies across the working plane, so an average figure is used.

- *Service illuminance* **is the mean illuminance achieved during the maintenance cycle of a lighting system, and averaged over the area being considered**

The illuminance needed for a particular task depends upon the visual difficulty of the task, the average standard of eyesight involved, and the type of performance expected. It can be shown that the speed and accuracy of various types of work are affected by the level of illuminance supplied.

Table 6.3 gives some of the standard service illuminance levels that are recommended for a variety of interiors and tasks. The values represent current good practice, which takes into account visual needs, practical experience, and efficient use of energy relating to the situation.

Lumen method

Chapter 5 explained how illuminance is affected by the distance and the angles between the illuminated surface and the light source, and by the reflectances of the surrounding surfaces. In an interior where there is more than one light source and there are several reflecting surfaces, the repeated combination of these effects makes calculation difficult if basic formulas are used. Some simplified methods have been developed and are found to give satisfactory results.

The lumen method is a commonly used technique of lighting design which is valid if the luminaires are mounted overhead in a regular pattern. The luminous flux output (lumens) of each lamp needs to be known, as do details of the luminaires and the room surfaces. Usually the illuminance level is already specified, the designer chooses suitable lamps and luminaires, and then wishes to know how many fittings are required to meet the specification. The number of lamps is given by the following formula:

$$N = \frac{E \times A}{F \times UF \times LLF}$$

where $N =$ number of lamp fittings required
$E =$ illuminance level required (lux)
$A =$ area at working plane height (m^2)
$F =$ initial luminous flux output of each lamp (lm)
$UF =$ Utilisation Factor – an allowance for the distribution effects of the luminaire and the room surfaces, as described below
$LLF =$ Light Loss Factor – an allowance for a reduction in light output caused by lamp deterioration and dirt. Details are given below.

Utilisation Factor
- **The *Utilisation Factor* is the ratio of the total flux reaching the working plane compared with the total flux output of the lamps**

Table 6.3 *Standard service illuminances*

Location	Illuminance (lux)	Position
General areas		
Entrance halls	150	1.2 m
Stairs	150	treads
Passageways	100	1.2 m
Outdoor entrances	30	ground
General assembly		
Casual work	200	wp
Rough work (e.g. heavy machinery)	300	wp
Medium work (e.g. vehicle bodies)	500	wp
Fine work (e.g. electronic assembly)	1000	bench
Very fine work (e.g. instrument assembly)	1500	bench
Offices		
General clerical	500	desk
Typing room	750	copy
Drawing offices	750	boards
Filing rooms	300	labels
Shops		
Counters	500	horizontal
Supermarkets	500	vertical
Education		
Chalkboard	500	vertical
Classrooms	300	desk
Laboratories	500	bench
Hotels		
Bars	150	table
Restaurants	100	table
Kitchens	500	wp
Homes		
General living room	50	wp
Casual reading	150	task
Studies	300	task
Kitchen	300	wp
General bedroom	50	floor
Halls and landings	150	floor
Recreation		
Gymnasium	500	floor
Squash rackets	300	floor
Swimming pool	300	water
Table tennis	500	table

Notes
Values from *CIBSE Code of Interior Lighting*.
wp: working plane.

Table 6.4 *Utilisation Factors for some luminaires*

Description of fitting	Basic downward LOR %	Room index	Ceiling 0.7 Walls 0.5	0.3	0.1	Ceiling 0.5 Walls 0.5	0.3	0.1	Ceiling 0.3 Walls 0.5	0.3	0.1
Aluminium industrial reflector, Aluminium or enamel high-bay reflector	70	0.6	0.39	0.36	0.33	0.39	0.36	0.33	0.39	0.35	0.33
		0.8	0.48	0.43	0.40	0.46	0.43	0.40	0.46	0.43	0.40
		1.0	0.52	0.49	0.45	0.52	0.48	0.45	0.52	0.48	0.45
		1.25	0.56	0.53	0.50	0.56	0.53	0.49	0.56	0.52	0.42
		1.5	0.60	0.57	0.54	0.59	0.57	0.53	0.59	0.55	0.53
		2.0	0.65	0.62	0.59	0.63	0.60	0.58	0.63	0.59	0.57
		2.5	0.67	0.64	0.62	0.65	0.62	0.61	0.65	0.62	0.60
		3.0	0.69	0.66	0.64	0.67	0.64	0.63	0.67	0.64	0.62
		4.0	0.71	0.68	0.67	0.69	0.67	0.65	0.69	0.66	0.64
		5.0	0.72	0.70	0.69	0.71	0.69	0.67	0.71	0.67	0.66
Near-spherical diffuser, open beneath	50	0.6	0.28	0.22	0.18	0.25	0.20	0.17	0.22	0.18	0.16
		0.8	0.39	0.30	0.26	0.33	0.28	0.23	0.27	0.25	0.22
		1.0	0.43	0.36	0.32	0.38	0.34	0.29	0.31	0.29	0.26
		1.25	0.48	0.41	0.37	0.42	0.38	0.33	0.34	0.32	0.29
		1.5	0.52	0.46	0.41	0.46	0.41	0.37	0.37	0.35	0.32
		2.0	0.58	0.52	0.47	0.50	0.48	0.43	0.42	0.39	0.36
		2.5	0.62	0.56	0.52	0.54	0.50	0.47	0.45	0.42	0.40
		3.0	0.65	0.60	0.56	0.57	0.53	0.50	0.48	0.45	0.43
		4.0	0.68	0.64	0.61	0.60	0.56	0.54	0.51	0.48	0.46
		5.0	0.71	0.60	0.65	0.62	0.59	0.57	0.53	0.50	0.48
Recessed louvre trough with optically designed reflecting surfaces	50	0.6	0.28	0.25	0.23	0.28	0.25	0.23	0.28	0.25	0.23
		0.8	0.34	0.31	0.28	0.33	0.30	0.28	0.33	0.30	0.28
		1.0	0.37	0.36	0.32	0.37	0.34	0.32	0.37	0.34	0.32
		1.25	0.40	0.38	0.35	0.40	0.37	0.35	0.40	0.37	0.35
		1.5	0.43	0.41	0.38	0.42	0.40	0.38	0.42	0.39	0.38
		2.0	0.46	0.44	0.42	0.45	0.43	0.41	0.44	0.42	0.41
		2.5	0.48	0.46	0.44	0.47	0.45	0.43	0.46	0.44	0.43
		3.0	0.49	0.47	0.46	0.48	0.46	0.45	0.47	0.45	0.44
		4.0	0.50	0.49	0.48	0.49	0.48	0.47	0.48	0.47	0.46
		5.0	0.51	0.50	0.49	0.50	0.49	0.48	0.49	0.48	0.47

The Utilisation Factor can be calculated directly or, more usually, obtained from tables which combine the distribution properties of the luminaire with the room index and with the reflectances of the room surface. Table 6.4 lists the Utilisation Factors for some representative types of luminaire.

Light Loss Factor

- **The *Light Loss Factor* is the ratio of the illuminance provided at some given time compared with the initial illuminance**

Light Loss Factor is calculated as the product of the three other factors:

LLF = Lamp Maintenance Factor × Luminaire Maintenance Factor × Roof Surface Maintenance Factor

where the *Lamp Maintenance Factor* is an estimate of the decline in output of the lamp source over a set time

the *Luminaire Maintenance Factor* is an estimate of the reduction in light output caused by dirt on the luminaire over a set time

the *Room Surface Maintenance Factor* is an estimate of the effect of dirt deposited on the room surfaces over a set time.

A range of values for Light Loss Factors is given in table 6.5. The determination of particular values depends on the type of lamps in use, the cleanliness of air in the district and the building, and whether dirt can easily collect on fittings

The *Maintenance Factor*, previously used instead of the Light Loss Factor, took account of lighting loss caused by dirt but did not include deterioration of lamps, for which an average output was assumed.

Table 6.5 *Typical light loss factors*

12 month *LLF*	Direct lighting	Indirect lighting
Air-conditioned building	0.95	0.9
Dirty industrial area	0.7	0.35

Notes
Room is of average cleanliness, like an office.
Lamps run for average 8 hours per working day.

Layout

The number of lamps needed, as calculated by the lumen formula, usually needs to be rounded up to a convenient figure and the layout of the luminaires decided upon. In order that the illuminance provided does not fall below a minimum value, the fittings must be placed in a regular grid pattern and their spacing must not exceed certain distances. This maximum

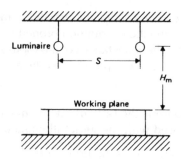

Figure 6.6 *Spacing of luminaires*

spacing depends on the type of luminaire and the height at which they are set. Typical values are as follows:

For fluorescent tubes in diffusing luminaires

$S_{max} = 1.5 \times H_m$

For filament lamps in direct luminaires

$S_{max} = 1.0 \times H_m$

where S_{max} = maximum horizontal spacing between fittings

H_m = mounted height of fitting above the working plane, as shown in figure 6.6. Take working plane as 0.85 m above floor level unless otherwise specified.

Lighting control

Switching off unwanted lights is an obvious method of saving energy and money. Building regulations require that some buildings achieve certain levels of energy saving by control of their lighting installations. Lighting control can be achieved by the following methods.

- **Timer control:** where timers are set to switch off lighting for periods of known inactivity, such as at the end of the working day.
- **Daylight control:** where lights are switched on or off, or dimmed, according to the level of daylight detected in a room.
- **Occupation control:** such as by sensors which detect noise or movement in an area. The sensors turn lighting on when there is someone in the area, and off again after a time delay if there is nobody in the room.
- **Local switching:** where it is possible to switch on lights only in the part of the room which is being occupied.

Worked example 6.2
A factory space measuring 40 m by 12 m by 4 m in height requires a service illuminance of 500 lux on the work benches which are set 1 m above the

floor. The 65 W tubular fluorescent lamps chosen have a luminous efficacy of 80 lm/W. They are to be mounted on the ceiling in luminaires which have a DLOR of 50 per cent. The room reflectances are 0.5 for the ceiling and 0.3 for the walls; the initial light loss factor is 0.7.

(a) Use the lumen method of design to calculate the number of lamps required.
(b) Suggest a suitable layout for the lamp fittings.

Know $E = 500 \, \text{lx}$, $\quad F = 65 \times 80 = 5200 \, \text{lm}$
$\qquad L = 40 \, \text{m}$, $\quad W = 12 \, \text{m}$, $\quad H_m = 4 - 1 = 3 \, \text{m}$
$\qquad A = 40 \times 12 = 480 \, \text{m}^2$, $\quad LLF = 0.7$

$$RI = \frac{L \times W}{H_m(L+W)}$$

$$= \frac{40 \times 12}{3(40+12)} = 3$$

Reflectances: ceiling 0.5, walls 0.3

$UF = 0.46$ (from table 6.4 using above data)

Using

$$N = \frac{E \times A}{F \times UF \times LLF}$$

$$= \frac{500 \times 480}{5200 \times 0.46 \times 0.7} = 143.3$$

So number of lamps required = **144 lamps**

Suggested layout: 9 rows of 16 luminaires

Check spacing using $S_{max} = 1.5 \times H_m$
$$= 1.5 \times 3 = 4.5 \, \text{m}$$

So the suggested layout is satisfactory provided that the distance between lamps is not greater than 4.5 m.

Glare index

Glare was defined in the previous chapter as the discomfort or impairment of vision caused by an excessive range of brightness in the visual field. The usual causes of glare in buildings are bright skies seen through windows and direct views of bright lamps. The glare is most likely to be discomfort glare which, over a period of time, can cause annoyance and affect efficiency.

- *GLARE INDEX* **is a numerical measure of discomfort glare which enables glare to be assessed and acceptable limits recommended**

The glare index is calculated from the following: a knowledge of the positions of the source and the viewpoint; the luminances of the source and the surroundings; and the size of the source. The calculated index may be compared with the maximum index recommended for the particular environment.

Glare indices for artificial lighting usually range from 10 for a low-brightness fitting, to 30 for an unshielded lamp. A typical glare index is 19, which is the recommended limit for offices.

Lighting criteria

Many factors may be relevant to the design of a particular lighting system. The following list summarises the main factors that usually need to be considered; the priorities will depend upon the type of situation.

- **Light quantity:** depends upon the nature of the task and the light output of lamp and luminaire. It is usually specified by illuminance level in lux.
- **Natural light:** may be used as a complete source of light or to supplement artificial light sources. Daylighting is the topic of the next chapter.
- **Colour quality:** depends upon the requirements of the task and the colour rendering properties of the source. Methods of specifying colour quality include spectral distribution, colour temperature, and the colour rendering index.
- **Glare:** depends upon the brightness and contrast of light sources and surfaces, and the viewing angles. It is usually specified by a glare index.
- **Directional quality:** depends upon the three-dimensional effect required, and the nature of the lamp and luminaire. It can be specified by vector and scalar luminance.
- **Energy use:** depends upon the electrical efficiency of the lamps and the use of switches. All lamps give off heat and are a source of heat gain in a building. Windows providing natural light can also be a significant source of heat loss and solar gain.
- **Costs:** depend upon the initial cost of the fittings; the cost of replacing the lamps (including labour); and the electricity consumption of the lamps.
- **Physical properties:** include size, appearance and durability of fittings.

Exercises

1 A certain space needs to be illuminated with a total luminous flux of 18 000 lm. The tungsten filament lamps used are rated at 60 W each and

have a luminous efficacy of 12 lm/W.

(a) Calculate the number of such lamps required.

(b) Calculate the number of kilowatt hours of electrical energy used by the lamps in a 12 hour period. (1 kWh = 1 kW × 1 hour.)

2 The level of illumination described in exercise 1 is required 12 hours a day for 2 years. Compare the differences in initial costs, replacement costs, and the energy costs using the following types of lamps:

(a) compact fluorescent lamps; and

(b) tungsten filament lamps.

Use manufacturers' data or your own estimates for costs of lamps, fittings, and electrical energy. Table 6.1 may also be useful.

3 An area measuring 18 m by 8 m is to have a service illuminance of 300 lx. The tubular fluorescent lamps each have a luminous flux output of 2820 lm and the luminaires give a utilisation factor of 0.4. The Light Loss Factor assumed is 0.8. Calculate the number of lamps required and suggest a layout for them.

4 A classroom with an area 10 m by 6 m is illuminated by 18 tubular fluorescent lamps, where each lamp is 60 W with a luminous efficacy of 80 lm/W. The Utilisation Factor is 0.46 and the Light Loss Factor used is 0.8. Calculate the average illuminance in the classroom.

5 A workshop is 12 m by 6 m by 4 m high and has workbenches 1 m high. Discharge lamps, each with an output of 3700 lm, are to be fitted in aluminium industrial reflectors at ceiling level. The surfaces have reflectances of 0.7 for the ceiling and 0.5 for the walls. The light loss factor is 0.7. The illuminance required on the workbenches is 400 lx.

(a) Find the Utilisation Factor for the room (use table 6.4).

(b) Calculate the number of lamps required.

Answers

1 (a)   25 lamps; **(b)** 18 kWh

3 48 lamps

4 530 lx

5 (a) 0.56; **(b)** 20 lamps

7 Natural Lighting

It is often necessary to provide a room with natural light from the Sun or the sky. The qualities of this natural light may be thought desirable for a pleasant environment or they may be needed to perform certain tasks, such as exacting work with colour. The natural light can be used as the sole source of interior lighting or can be combined with artificial light.

Daylight is usually admitted into a building by means of windows or skylights; but these windows also transmit heat, sound and perhaps air. So the design of windows for a building, called *fenestration*, affects almost all the environmental variables. The provision of natural lighting in a building must not be designed without also considering questions of artificial lighting, heating, ventilation and sound control.

The quantity of natural light inside a room is governed by the factors listed below. By analysing these factors it is possible to describe daylight numerically and to predict its effects in a room.

- The nature and brightness of the sky
- The size, shape and position of the windows
- Reflections from surfaces inside the room
- Reflections and obstructions from objects outside the room.

NATURAL LIGHT SOURCES

All natural daylight comes from the Sun, by way of the sky. But, while the Sun can be considered as a reasonably constant source of light, the light from the sky varies with the time of the day, with the season of the year and with the local weather.

In parts of the world with predominantly dry, sunny weather much of the natural light inside buildings is direct sunlight which has been reflected. In the United Kingdom, and similar countries where sunshine is unreliable, the overcast sky is considered as the main source of natural light. Because the sky continually varies it is necessary to define certain 'standard skies', with constant properties, for use in design work.

160

Direct sunlight

The levels of illuminance on the ground provided by light direct from the Sun may be as high as 100 000 lx in the summer. Direct sunlight should generally be avoided inside working buildings because it can easily cause unacceptable glare. Sun control devices were discussed in chapter 3. However, in domestic buildings a certain amount of Sun penetration is considered desirable by most people who live in temperate climate zones, such as in North-West Europe. One guideline for the UK is that sides of all residential buildings which face east, south and west should have at least one hour of sunshine on 1 March.

Uniform standard sky

The uniform sky is a standard overcast sky which is taken to have the same luminance in every direction of view. Viewed from an unobstructed point on the ground the sky effectively makes a hemispherical surface around that point. This surface emits light and the uniform sky assumes that every part of the surface is equally bright.

The illuminance at a point on the ground provided by an unobstructed overcast sky continually varies but a constant 5000 lx is taken as one standard for daylight calculations in Britain. This represents the light from a heavily overcast sky and is a conservative design assumption – the 5000 lx is actually exceeded for 85 per cent of normal working hours throughout the year.

Although the overcast sky of countries like the UK is not strictly uniform in luminance the model is useful for some purposes. A sky of uniform luminance is also a reasonable description of the clear skies in sunny climates.

CIE standard sky

The CIE sky is a standard overcast sky in which the luminance steadily increases above the horizon. This sky was defined by the Commission Internationale d'Eclairage as being described by the following formula and illustrated in figure 7.1:

$$L_\theta = L_z \, 1/3\left(1 + 2\sin\theta\right)$$

where L_θ = luminance of the sky at an altitude θ degrees above the horizon
L_z = luminance of the sky at the zenith.

The luminance of the CIE sky at the zenith is *three* times brighter than at the horizon. The CIE model is found to be a good representation of conditions in many parts of the world, especially in regions like North-West Europe.

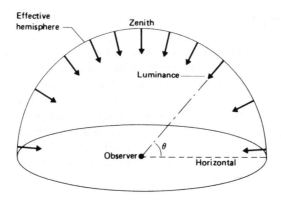

Figure 7.1 *Sky hemisphere*

DAYLIGHT FACTORS

The natural light that provides illumination inside a room is usually only a small fraction of the total light available from a complete sky. The level of illuminance provided by the sky varies as the brightness of the sky varies so it is not possible to specify daylight by a fixed illuminance level in lux.

The amount of daylight inside a room can be measured by comparing it with the total daylight available outside the room. This ratio, or daylight factor, remains constant for a particular situation because the two parts of the ratio vary in the same manner as the sky changes.

- ***Daylight Factor* is the ratio between the actual illuminance at a point inside a room and the illuminance possible from an unobstructed hemisphere of the same sky**

  UNIT: percentage

Direct sunlight is excluded from both values of illuminance, and the daylight factor can be expressed by the following formula:

$$DF = \frac{E_i}{E_o} \times 100$$

where DF = daylight factor at a chosen reference point in the room (per cent)

E_i = illuminance at the reference point (lx)

E_o = illuminance at that point *if* the sky was unobstructed (lx).

The definition is a theoretical one as it is not possible to measure instantly both types of illuminance at the same place. For purposes of specification and design a standard sky is assumed to give a minimum level of illuminance on the ground, and 5000 lx is a commonly-used value.

Worked example 7.1

A minimum daylight factor of 4 per cent is required at a certain point inside a room. Calculate the natural illuminance that this represents, assuming that an unobstructed standard sky gives an illuminance of 5000 lx.

Know $DF = 4$ per cent, $E_o = 5000\,lx$, $E_i = ?$

Using

$$DF = \frac{E_i}{E_o} \times 100$$

$$4 = \frac{E_i}{5000} \times 100$$

$$E_i = \frac{4 \times 5000}{100} = 200$$

So illuminance = **200 lx**

Recommended daylight factors

Daylight factors can be used to specify recommended levels of daylight for various interiors and tasks. Table 7.1 lists a selection of recommendations for interiors where daylight from side windows is a major source of light. Daylight factors vary for different points within a room so it is usual to quote average values or minimum values.

Table 7.1 *Typical daylight levels*

Location	Average daylight factor	Minimum daylight factor	Surface
General office	5	2	desks
Classroom	5	2	desks
Entrance hall	2	0.6	working plane
Library	5	1.5	tables
Drawing office	5	2.5	boards
Sports hall	5	3.5	working plane

Daylight factor components

The daylight reaching a particular point inside a room is made up of three principal components. The three components arrive at the same point by different types of path, as indicated in figure 7.2.

1. **Sky component (SC):** the light received directly from the sky.
2. **Externally reflected component (ERC):** the light received directly by reflection from buildings and landscape outside the room.
3. **Internally reflected component (IRC):** the light received from surfaces inside the room.

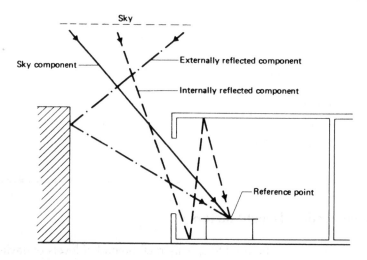

Figure 7.2 *Components of daylight factor*

The use of daylight factor components is a step towards predicting the daylight factors at a point. The three components can be analysed and calculated separately. The final daylight factor is the sum of the three separate components, so that:

- $DF = SC + ERC + IRC$

where DF = Daylight factor
        SC = Sky component
      ERC = Externally reflected component
       IRC = Internally reflected component.

Daylight factor contours

Daylight factor contours represent the distribution of daylight inside a room by means of lines which join points of equal daylight factor. The contours are commonly shown on a plan of the room at working plane height, as shown in figure 7.3. Contours around windows have characteristic lobe

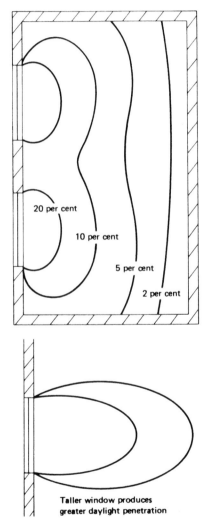

Figure 7.3 *Daylight factor contours*

shapes which converge at the edge of the windows. Tall windows provide greater penetration of contours and multiple windows cause contours to join.

Assessment of daylight factor

When a building already exists the values and distribution of daylight inside the building can be measured directly. A *daylight factor meter* is a specially calibrated light meter (photometer) which gives direct readings of the daylight factor at any point.

The prediction of daylight factors at the design stage requires a knowledge of the proposed building and its surroundings. It is possible to calculate the three components of the daylight factor by using information about the size of the windows and room, the size of any external obstructions, and the proposed reflectances of the surfaces. The sky component is the major contributor to a daylight factor and can be considered as the percentage of an unobstructed sky that is visible from the reference point.

The following methods can be used for predicting the daylight factor in a building:

- Tables of window and room dimensions
- Grids of the sky such as the Waldram diagram
- Computer programs
- Physical models measured in an artificial sky room
- Daylight factor protractors.

The use of special tables, diagrams, and protractors are techniques which help bypass the many repetitive calculations needed to predict the daylight factor at each point within a room. The *Waldram diagram*, for example, is a specially scaled grid representing half the hemisphere of sky. Using scaled plans of the room the area of sky visible through the window from the reference point is plotted onto the grid. The area of grid covered by this plot is proportional to the sky component at the reference point.

Computers are easily programmed to repeatedly make the tedious calculations needed for the prediction of daylight factor. Modern software packages for the prediction of daylight ask you to enter the details of your room, windows and reflecting surfaces, and to specify a grid pattern within the room. After calculating the daylight factor expected at each grid point the results may be shown on screen as daylight factor contours between the points and also given as a printout.

BRE daylight factor protractors

The special protractors developed by the Building Research Establishment are widely used to determine daylight factors at the design stage of a

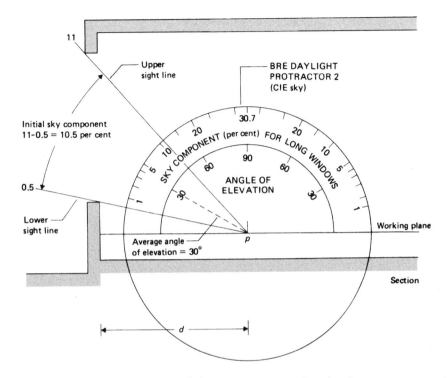

Figure 7.4 *BRE Daylight Protractor on section drawing*

building. Protractors are available for glazing set at different angles and for either a uniform sky or a CIE sky.

A daylight protractor, as shown in figure 7.4, contains two semi-circular scales on transparent overlays, which are used with scale drawings of the room being assessed. The primary scale measures the initial sky component and an auxiliary scale makes a correction for the width of the windows. The protractor also contains an ordinary scale of angle used to find the angle of elevation of the patch of sky being assessed.

Calculation of sky component

Step 1: Take a section and a plan drawing of the room, drawn to any scale. Mark reference points on the drawings, usually a regular grid of points at working plane height. Choose a protractor that is suitable for the angle of glazing and for the type of sky required.

Step 2: On the section draw sight lines from the chosen reference point to the top and bottom edges of sky visible from that point. Place the primary scale of the protractor over the section, aligned on the reference

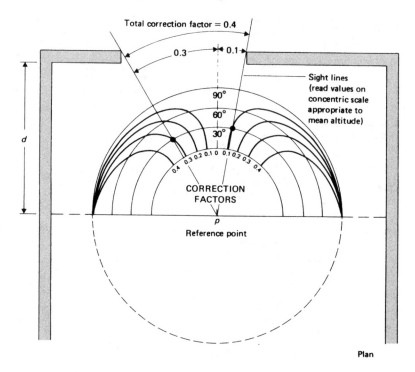

Figure 7.5 *BRE Daylight Protractor on plan drawing*

point, and read the two values of the sight lines, as shown in figure 7.5. Subtract the two readings to obtain the initial uncorrected value of sky component.

Step 3: Use the normal protractor scale to read the angles of the sight lines and average these readings to obtain the mean angle of elevation of the sky.

Step 4: On the plan draw sight lines from the same reference point to the vertical edges of the sky visible from that point. Place the auxiliary scale of the protractor over the plan, aligned on the reference point as shown in figure 7.5. Use the average angle of elevation to select the appropriate semicircular scale and read the values of the sight lines. To get a final correction factor add these readings if they are on opposite sides of the vertical; subtract the readings if they are both on the same side.

Step 5: The sky component for that point is equal to the initial value found in step 2 multiplied by the correction factor found in step 4.

Calculation of externally reflected component

The externally reflected component is initially calculated in the same manner as the sky component. Sight lines are drawn to the top edge of obstructions visible from the reference point. This gives an equivalent sky component, which is converted to the externally reflected component by allowing for the reduced luminance of the obstructing surface. For uniform sky protractors the initial ERC is divided by 10. For CIE sky protractors the initial ERC is divided by 5.

Calculation of internally reflected component

The internally reflected component of daylight inside a room depends on the reflectances of the room surfaces and on the size of the windows and obstructions. The process of multiple reflections is complex but formulas and tables are available to calculate the IRC. Table 7.2 gives a simplified example.

Table 7.2 *Internally reflected components*

Window area as % of floor area	Floor reflection factor							
	(10%)				(40%)			
	Wall reflection factor							
	(20%)	(40%)	(60%)	(80%)	(20%)	(40%)	(60%)	(80%)
5	0.1	0.1	0.2	0.4	0.1	0.2	0.4	0.6
10	0.1	0.2	0.4	0.7	0.3	0.5	0.8	1.2
20	0.2	0.5	0.8	1.4	0.5	0.9	1.5	2.3
30	0.3	0.7	1.2	2.0	0.8	1.3	2.1	3.3
40	0.5	0.9	1.6	2.6	1.0	1.7	2.7	4.2
50	0.6	1.1	1.9	3.1	1.3	2.1	3.2	4.9

This table gives the minimum IRC values for rooms of approx. 40 m^2 floor area, 3 m height, 70%, reflection factor and with window on one side.
Adapted from data in *BRE Digests*.

Worked example 7.2

Calculate the sky component for the room and reference point illustrated in figures 7.4 and 7.5. Use the readings shown on the protractors.

Step 1: The drawings of the room and the reference point are shown in figures 7.4 and 7.5. The protractor is type number 2, for vertical glazing and a CIE sky.

Step 2: From primary scale:

top reading	11
subtract bottom reading	0.5
Uncorrected sky component	0.5 per cent

Step 3: From elevation scale:

top reading	50°
bottom reading	10°
Average angle of elevation	60/2 = 30°

Step 4: From auxiliary scale:

first reading	0.3
second reading	0.1
Correction factor	0.4

Step 5

Sky component = Uncorrected sky component × correction factor
= 10.5 × 0.4 = 4.2

So sky component = **4.2 per cent**

COMBINED LIGHTING

People prefer to work by natural light but it is difficult to provide adequate levels of natural illumination to all parts of an interior. Areas that are quite close to windows may still require extra illumination for certain purposes and artificial lighting then needs to be combined with the natural lighting. This system of combined lighting requires careful design so as to preserve the effect of daylight as much as possible, and to make the best use of energy.

PSALI

PSALI is an abbreviation for *Permanent Supplementary Artificial Lighting of Interiors.*

PSALI is a system of combined daylighting and artificial lighting where parts of an interior are lit for the whole time by artificial light, which is designed to balance and blend with the daylight

The use of PSALI retains most of the psychological benefits of full daylighting but allows the use of deeper room plans, which can save energy because of lower heat losses. The guiding principle of PSALI is to provide illumination that appears to be of good daylight character, even though most of the working illumination might be supplied by unobtrusive artificial light sources.

To achieve an appropriate blend of lighting the following factors should be considered.

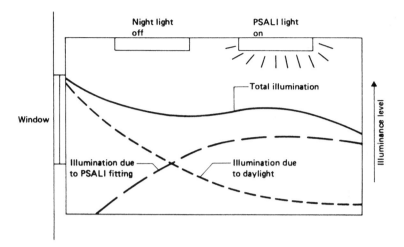

Figure 7.6 *Example of PSALI system*

Distribution of light

The total illuminance should gradually increase towards the windows. The illumination to be provided by PSALI can be determined by choosing a final illumination curve, as shown in figure 7.6, and then subtracting average daylight values. The illuminance over the main working areas should not vary by more than about 3 to 1.

Sudden changes in luminance between room surfaces should be avoided. In general this requires neutral colour schemes, which have the same appearance under both natural and artificial light.

Choice of lamps

The lamps chosen for the artificial lighting should match the natural light in colour appearance. Daylight is variable in colour quality, with a different spectral output to any lamp, so a compromise must be made. Tubular fluorescent lamps with colour temperatures in the range 4000–6500 K are usually employed for PSALI. The fittings should be unobtrusive and are often recessed in the ceiling.

Switching

A combined lighting system requires several lighting control circuits – for the PSALI during daylight hours and for the complete artificial lighting after dark. Figure 7.6 indicates a simple system where extra lights would be switched on at night or during dull days. Some types of building can save electrical energy by controlling the artificial lighting with photo-electric cells which sense changing daylight levels and switch lights on and off as necessary.

Exercises

1 Calculate the luminance of a CIE standard sky at an altitude of 20° if the luminance at the zenith is taken to be 2200 cd/m².

2 The natural illuminance at a point inside a room is 430 lx and the illuminance given by an unobstructed sky is assumed to be 5380 lx. Calculate the daylight factor at that point.

3 Draw a plan of a convenient classroom or office and sketch the shapes of the daylight factors that would be expected from the windows. If possible, investigate the actual distribution of daylight with a light meter and/or computer program.

Answers

1 1235 cd/m²

2 8 per cent

8 Principles of Sound

Good building design invariably involves a consideration of the presence of sound in the environment. Common topics of concern are the exclusion of external noise, the reduction of sound passing between rooms, and the quality of sound inside rooms. Before these topics are studied, this chapter outlines the basic principles of sound and its measurement.

NATURE OF SOUND

Origin of sound

Sound is a sensation produced in the ear and brain by variations in the pressure of the air. These pressure variations transfer energy from a source of vibration. Suitable vibrations in the air can be caused by a variety of methods, such as those indicated below:

- **Moving objects:** Examples include loudspeakers, guitar strings, vibrating walls, and human vocal chords.
- **Moving air:** Examples include horns, organ pipes, mechanical fans, and jet engines.

A vibrating object compresses adjacent particles of air as it moves in one direction and leaves the particles of air 'spread out' as it moves in the other direction. This effect is illustrated in figure 8.1. The displaced particles pass on their extra energy and a pattern of compressions and rarefactions travels out from the source, while the individual particles return to their original positions.

Wave motion

Although the individual particles of air return to their original position the sound *energy* obviously travels forward and does so in the form of a wave motion. The front of the wave spreads out equally in all directions unless it is affected by an object or by another material in its path.

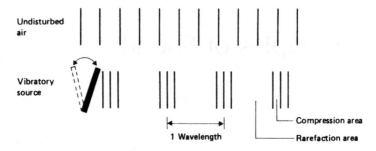

Figure 8.1 *Pressure variations in a sound wave*

The waves are *longitudinal* in type because the particles of the medium carrying the wave vibrate in the same direction as the travel of the wave, as shown in figure 8.1. The sound waves can travel through solids, liquids, and gases, but not through a vacuum.

It is difficult to depict a longitudinal wave in a diagram so it is convenient to represent the wave as shown in figure 8.2, which is a plot of the vibrations against time. For a pure sound of one frequency, as in figure 8.2, the plot takes the form of a sine wave.

Sound waves are like any other wave motion and therefore can be specified in terms of wavelength, frequency, and velocity.

- **WAVELENGTH (λ) is the distance between any two repeating points on a wave**

  UNIT: metre (m)

In figure 8.1 a wavelength is shown measured between two compressions but the length between any two repeating points would be the same.

- **FREQUENCY (f) is the number of cycles of vibration per second**

  UNIT: hertz (Hz)

Figure 8.2 shows two complete vibrations, or cycles.

- **VELOCITY (v) is the distance moved per second in a fixed direction**

  UNIT: metres per second (m/s)

For every vibration of the sound source the wave moves forward by one wavelength. The number of vibrations per second therefore indicates the total length moved in 1 second; which is the same as velocity. This relationship is true for all wave motions and can be written as the following formula:

$$v = f \times \lambda$$

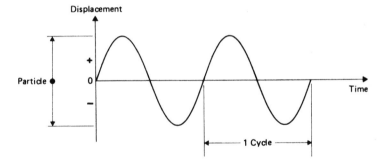

Figure 8.2 *Vibrations of a sound wave*

where v = velocity in m/s
f = frequency in Hz
λ = wavelength in m.

Worked example 8.1
A particular sound wave has a frequency of 440 Hz and a velocity of 340 m/s. Calculate the wavelength of this sound.

Know v = 340 m/s, f = 440 Hz, λ = ?

Using

$$v = f \times \lambda$$
$$340 = 440 \times \lambda$$

$$\lambda = \frac{340}{440} = 0.7727$$

So wavelength = **0.7727 m**

Velocity of sound

A sound wave travels from its source with a steady velocity that is independent of the rate at which the vibrations occur. This means that the frequency of a sound does not affect its speed.

The velocity of sound is affected by the properties of the material through which it is travelling. The velocity in air increases as the temperature or humidity increases. The velocity is, however, unaffected by variations in atmospheric pressure such as those caused by the weather. An indication of the velocities of sound in different materials is given in table 8.1.

Table 8.1 *Velocity of sound*

Material	Typical velocity (m/s)
Air (0 °C)	331
Air (20 °C)	344
Water (25 °C)	1498
Pine	3300
Glass	5000
Iron	5000
Granite	6000

Sound travels faster in liquids and solids than it does in air because the densities and elasticities of those materials are greater. The particles of such materials respond to vibrations more quickly and so convey the pressure vibrations at a faster rate.

Frequency of sound

If an object that produces sound waves vibrates 100 times a second, for example, then the frequency of that sound wave will be 100 Hz. The human ear hears this as sound of a certain pitch.

- *PITCH* **is the frequency of a sound as perceived by human hearing**

Low-pitched notes are caused by low-frequency sound waves and high-pitched notes are caused by high-frequency waves. The pitch of a note determines its position in the musical scale. The frequency range to which the human ear responds is approximately 20 to 20 000 Hz and frequencies of some typical sounds are shown in figure 8.3.

Most sounds contain a combination of many different frequencies and it is usually convenient to measure and analyse them in ranges of frequencies, such as the octave.

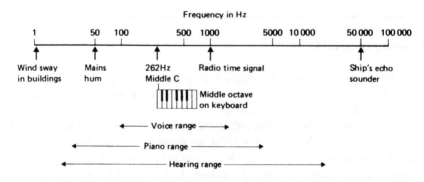

Figure 8.3 *Frequency ranges of sound waves*

- **An *OCTAVE BAND* is the range of frequencies between any one frequency and double that frequency**

For example, 880 Hz is one octave above 440 Hz. Octave bands commonly used in frequency analysis have the following centre frequencies:

31.5   63   125   250   500   1000   2000   4000   8000 Hz

The upper frequency of an octave band can be found by multiplying the centre frequency by a factor of $\sqrt{2}$.

Quality of sound

A pure tone is sound of only one frequency, such as that given by a tuning fork or electronic signal generator. Most sounds heard in everyday life are a mixture of more than one frequency, although a lowest *fundamental* frequency predominates when a particular 'note' is recognisable. This fundamental frequency is accompanied by *overtones* or *harmonics.*

- ***OVERTONES* and *HARMONICS* are frequencies equal to whole-number multiples of the fundamental frequency**

For example, the initial overtones of the note with a fundamental of 440 Hertz are as follows:

440 Hz = fundamental or 1st harmonic
880 Hz = 1st overtone or 2nd harmonic
1320 Hz = 2nd overtone or 3rd harmonic etc.

Different voices and instruments are recognised as having a different quality when making the same note. This individual *timbre* results because different instruments produce different mixtures of overtones which accompany the fundamental, as shown in figure 8.4. The frequencies of these overtones may well rise to 10 000 Hz or more and their presence is often an important factor in the overall effect of a sound. A telephone, for example, transmits few frequencies above 3000 Hz and the exclusion of the higher overtones noticeably affects reproduction of the voice and of music.

Resonance

Every object has a *natural frequency* which is the characteristic frequency at which it tends to vibrate when disturbed. For example, the sound of a metal bar dropped on the floor can be distinguished from a block of wood dropped in the same way. The natural frequency depends upon factors such as the shape, density, and stiffness of the object.

Resonance occurs when the natural frequency of an object coincides with the frequency of any vibrations applied to the object. The result of resonance is extra large vibrations at this frequency.

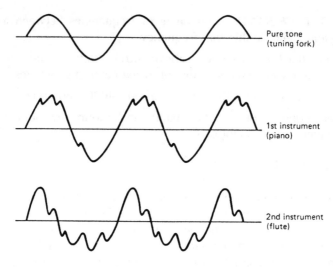

Figure 8.4 *Different waveforms with different overtones*

Resonance may occur in many mechanical systems. For instance, it can cause loose parts of a car to rattle at certain speeds when they resonate with the engine vibrations. The shattering of a drinking glass has been attributed to resonance of the object with a singer's top note! Less dramatic, but of practical application in buildings, is that resonance affects the transmission and absorption of sound within partitions and cavities.

SOUND LEVELS

The strength or 'loudness' of a sound depends upon its energy content and this energy affects the size of the pressure variations produced. The amplitude of the sound wave, the maximum displacement of each air particle, is greater for stronger sounds as shown in figure 8.5. Notice that the frequency or note of the sound remains unaltered.

Measurement

To specify the strength of a sound it is usually easiest to measure or describe some aspect of its energy or its pressure. Even so, sound does not involve large amounts of energy and its effect depends upon the high sensitivity of our hearing.

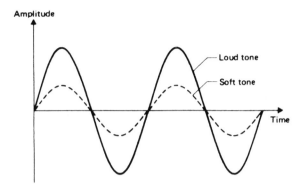

Figure 8.5 *Waveforms of soft and loud sounds*

Sound power
- **SOUND POWER (P) is the rate at which sound energy is produced at the source**

  UNIT: watt (W)

Sound power is a fundamental property of a sound *source* but is difficult to measure directly. The maximum energy output of a voice is about 1 mW, which explains why talking does not usually exhaust us! In terms of sound power a typical jet engine produces only several kilowatts.

Sound intensity
- **SOUND INTENSITY (I) is the sound power distributed over unit area**

  UNIT: watts per square metre (W/m^2)

Sound intensity is a measure of the rate at which energy is received at a given surface. If, for example, a source is radiating sound in all directions then the sound spreads out in the shape of a sphere and the intensity is found from the following formula:

$$I = \frac{P}{4\pi r^2}$$

where I = intensity at distance (W/m^2)
P = sound power of the source (W)
r = distance from the source (m).

Sound pressure
- **SOUND PRESSURE (p) is the average variation in atmospheric pressure caused by the sound**

  UNIT: pascal (Pa)

The pressure is continuously varying between positive and negative values so it is measured by its *root mean square* (RMS) value, which is a type of average having positive values only.

- RMS pressure = 0.707 maximum pressure

Intensity–pressure relationship

The intensity of a sound is proportional to the square of its pressure and expressed by the following formula:

$$I = \frac{p^2}{\varrho v}$$

where I = intensity of the sound (W/m²)
   p = pressure of the sound (Pa)
   ϱ = density of the material (kg/m³)
   v = velocity of sound (m/s).

Thresholds

The weakest sound that the average human ear can detect is remarkably low and occurs when the membrane in the ear is deflected by a distance less than the diameter of a single atom.

Threshold of hearing

- **The *THRESHOLD OF HEARING* is the weakest sound that the average human ear can detect**

The value of the threshold varies slightly from person to person but for reference purposes it is defined to have the following values at 1000 Hz:

$I_0 = 1 \times 10^{-12}$ W/m² when measured as intensity
$p_0 = 20 \times 10^{-6}$ Pa when measured as pressure

Threshold of pain

- **The *THRESHOLD OF PAIN* is the strongest sound that the human ear can tolerate**

Very strong sounds become painful to the ear. Excessive sound energy will damage the ear mechanism and very large pressure waves will have other harmful physical effects, such as those experienced in an explosion. The threshold of pain has the following approximate values:

$I = 100$ W/m²   or $p = 200$ Pa

Decibels

Although it is quite accurate to specify the strength of a sound by an absolute intensity or pressure, it is usually inconvenient to do so. For instance, the range of values between the threshold of hearing and the threshold of pain is a large one and involves awkward numbers. It is also found that, for the same change in intensity or pressure, the ear hears different effects when listening at high intensities and at low intensities. The ear judges differences in sound by ratios so that, for example, the difference between 1 and 2 Pa pressure is perceived to be the same difference as between 5 and 10 Pa.

For practical measurements of sound strength it is convenient to use a decibel scale based on constant ratios, a scale which is also used in some electrical measurements.

- **The *DECIBEL* (*dB*) is a logarithmic ratio of two quantities**

The decibel is calculated by the following formulas using either values of sound intensity or sound pressure. Velocity is proportional to the pressure squared:

$$N = 10\log_{10}\left(\frac{I_2}{I_1}\right) = 10\log_{10}\left(\frac{p_2}{p_1}\right)^2$$

where N = number of decibels

I_1 and I_2 are the two intensities being compared

or

p_1 and p_2 are the two pressures being compared.

Sound levels

In the measurement of sound levels the decibel ratio is always made with reference to the standard value for the threshold of hearing. This produces a scale of numbers that is convenient and gives a reasonable correspondence to the way that the ear compares sounds.

Figure 8.6 shows the total range of sound levels in decibels between the two thresholds of hearing and gives typical decibel values of some common sounds. Precise values would depend upon the frequencies contained in the sounds and the distances from the source.

The smallest change in sound level that the human ear can detect is 1 dB, although a 3 dB change is considered the smallest difference that is generally significant. A 10 dB increase or decrease makes a sound seem approximately twice as loud or half as loud, as shown in table 8.2.

Calculation of sound levels

Values of sound intensity or sound pressure are converted to decibels by comparing them with the standard value of the threshold of hearing. The word 'level' indicates that this reference has been used.

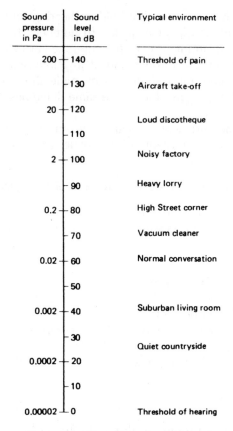

Figure 8.6 *Examples of sound levels*

Table 8.2 Changes in sound levels

Sound level change	Effect on hearing
±1 dB	negligible
±3 dB	just noticeable
+10 dB	twice as loud
−10 dB	half as loud
+20 dB	four times as loud
−10 dB	one quarter as loud

Sound intensity level

If the sound strength is considered in terms of intensity then a sound intensity level (SIL) is given by the formula

$$SIL = 10\log_{10}\left(\frac{I}{I_0}\right)$$

where I = the intensity of the sound being measured (W/m^2)

I_0 = the intensity of the threshold of hearing taken as $1 \times 10^{-12} W/m^2$.

Sound pressure level

Most practical instruments measure sound by responding to the sound pressure. The sound pressure level (SPL) is then given by the formula

$$SPL = 20\log_{10}\left(\frac{p}{p_0}\right)$$

where p = the RMS pressure of the sound being measured (Pa)

p_0 = the RMS pressure of the threshold of sound taken as $20\mu Pa$.

For most practical purposes, the SIL and the SPL give the same value in decibels for the same sound.

Worked example 8.2

A sound has a pressure of 4.5×10^{-2} Pa when measured under certain conditions. Calculate the sound pressure level of this sound. Threshold of hearing pressure = 20×10^{-6} Pa.

Know $p = 4.5 \times 10^{-2}$ Pa, $p_0 = 20 \times 10^{-6}$ Pa,   SPL = ?

Using

$$SPL = 20\log(p/p_0)$$

$$= 20\log\left(\frac{4.5\times10^{-2}}{20\times10^{-6}}\right) = 20\log(2250)$$

$$= 20\times3.3522 = 67.04$$

So SPL = **67 dB**

Worked example 8.3

Calculate the change in sound level when the intensity of a sound is doubled.

Let I = initial intensity, so $2I$ = final intensity.
Let L_1 = initial sound level and L_2 = final sound level.

$$L_2 - L_1 = 10\log\left(\frac{2I}{I_0}\right) - 10\log\left(\frac{I}{I_0}\right)$$

$$= 10\log\left(\frac{2I}{I_0} \div \frac{I_0}{I}\right) \text{ (by the rules of logarithms)}$$

$$= 10\log\frac{2I}{I} = 10\log 2$$

$$= 10 \times 0.3010 = 3.010$$

So change in SIL = **3 dB**
Therefore: doubling the energy gives a 3 dB increase in sound level.

Combination of sound levels

If two different sounds arrive at the same time then the ear is subject to two pressure waves. Because the decibel scale is logarithmic in origin the simple addition of sound levels in decibels does *not* give the sound level of the combined sounds. Two jet aircraft, for example, each with a SPL of 105 dB do not, fortunately, combine to give a total effect of 210 dB, which is well above the threshold of pain.

Although decibels cannot be directly added, intensities can be added, or the squares of pressures can be added, using the following formulas:

$$I = I_1 + I_2$$

or

$$P = \sqrt{\left(p_1^2 + p_2^2\right)}$$

When interpreting the results of combined sound levels, the following guidelines should be kept in mind:

- **3 dB** increase in sound level is caused by doubling the sound energy
- **10 dB** increase in sound level seems approximately twice as loud.

Worked example 8.4

Calculate the total sound level caused by the combination of sound levels of 95 dB and 90 dB. Threshold of hearing intensity = 1×10^{-12} W/m^2.

Let I_1 = intensity of 95 dB,
  I_2 = intensity of 90 dB, and
  I_3 = intensity of the combined sounds.

Using

$$SIL = 10\log\left(I/I_0\right)$$

$$95 = 10\log\left(\frac{I_1}{I_0}\right) \qquad \text{and} \qquad 90 = 10\log\left(\frac{I_2}{I_0}\right)$$

$$\log\left(\frac{I_1}{I_0}\right) = \frac{95}{10} \qquad\qquad \log\left(\frac{I_2}{I_0}\right) = \frac{90}{10}$$

$$\frac{I_1}{I_0} = \text{antilog } 9.5 \qquad\qquad \frac{I_2}{I_0} = \text{antilog } 9$$

$$\frac{I_1}{1 \times 10^{-12}} = 3.16 \times 10^9 \qquad\qquad \frac{I_2}{1 \times 10^{-12}} = 1 \times 10^9$$

$$I_1 = 3.16 \times 10^9 \times 10^{-12} \qquad\qquad I_2 = 1 \times 10^9 \times 10^{-12}$$

$$I_1 = 3.16 \times 10^{-3} \, \text{W/m}^2 \qquad\qquad I_2 = 1 \times 10^{-3} \, \text{W/m}^2$$

$$I_3 = I_1 + I_2$$

$$= \left(3.16 \times 10^{-3}\right) + \left(1 \times 10^{-3}\right) = 4.16 \times 10^{-3}$$

$$\text{Combined SIL} = 10\log\left(\frac{I_3}{I_0}\right) = 10\log\left(\frac{4.16 \times 10^{-3}}{1 \times 10^{-12}}\right)$$

$$= 10\log\left(4.16 \times 10^9\right)$$

$$= 10 \times 9.619$$

So total Sound level = **96 dB**

The addition of decibel values is made easier with the aid of a scale such as that shown in figure 8.7. The fact that the ear cannot detect differences less than 1 dB makes the inaccuracies of the table acceptable. From the scale for the addition of sound levels it can be seen that if the difference between two sound levels is greater than 15 dB then the addition of the lower level will produce negligible effect on the higher sound level.

These results mean that a significant sound, such as one of 80 dB, will not be heard above a similar type of sound at 95 dB. People can therefore be run

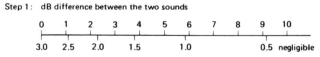

Step 1: dB difference between the two sounds

Step 2: dB correction added to higher level

Figure 8.7 *Addition of sound levels*

over by site vehicles because the sound of the vehicle is masked by nearby machinery noise. For the lower sound level to be noticed it should have a significantly different frequency content, such as that of a bell, siren, or telephone warbler.

ATTENUATION OF SOUND

As sound waves spread out from a source they attenuate – that is, their amplitude decreases and the sound level drops. Except for some absorption by the air the total energy of the wave front remains constant but the area of the wave front constantly increases. The energy therefore spreads over larger areas and the density of this energy (intensity) or the sound pressure measured at any point must decrease.

The manner in which the sound wave spreads and attenuates is affected by any directional effects of the source – a jet engine, for example, emits more noise to the rear than to the front. The propagation of the sound is also affected by any reflecting or blocking objects in the path.

A *free field* is one in which the sound waves encounter no objects. If there is an object in the sound path then some of the sound will be reflected, some will be absorbed, and some will be transmitted through the object. The exact effects depend on the nature of the object and the wavelength of the sound. In general, the size of the object must be greater than one wavelength of the sound wave in order to significantly affect the wave. A wavelength of one metre is typical of sound produced by voices.

For the initial prediction of the sound in the open air it is necessary to assume free field conditions and to consider the behaviour of sound being emitted from a point or a line type of source. These results may then be modified to take account of the reflective conditions encountered in most practical situations.

Point source of sound

In a free field the sound wave from a point source spreads out uniformly in all directions in the shape of a sphere, as shown in figure 8.8. The surface area of a sphere increases in proportion to the square of its radius and so the intensity of sound energy varies in an inverse manner.

Inverse square law
  - **The sound intensity from a point source of sound decreases in inverse proportion to the square of the distance from the source**

This relationship can also be expressed in the following equation:

$$I \propto \frac{1}{d^2}$$

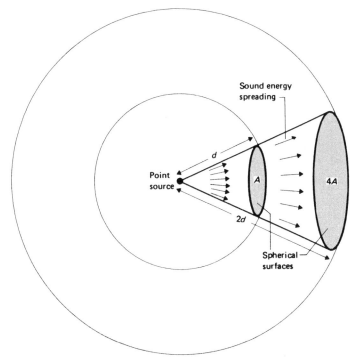

Surface area increases in proportion to the distance squared.
Intensity decreases in inverse proportion to the distance squared.

Figure 8.8 *Attenuation of sound from a point source*

where I is the sound intensity measured at distance d from the source.
    The ratio of any two intensities is given by the formula

$$\boxed{\frac{I_1}{I_2} = \frac{d_2^2}{d_1^2}}$$

where I_1 is the sound intensity measured at distance d_1 from the source and
    I_2 is the sound intensity measured at distance d_2 from the source.

Note: Observe the 'cross-over' within the formula because of the inverse
relationship.

Worked example 8.5
A microphone measures sound at a position in a free field 5 m from a point
source. Calculate the change in SPL if the microphone is moved to a position
10 m from the source.

Let L_1 = SPL at distance d_1 = 5 m
     L_2 = SPL at distance d_2 = 10 m

Using

$$L_1 - L_2 = 10\log(I_1/I_2) \quad \text{and} \quad I_1/I_2 = \left(d_2^2/d_1^2\right)$$

gives

$$L_1 - L_2 = 10\log\left(\frac{d_2^2}{d_1^2}\right) = 10\log\left(\frac{10^2}{5^2}\right)$$

$$= 10\log\left(\frac{100}{25}\right) = 10\log 4$$

$$= 10 \times 0.6021 = 6.021$$

So change in SPL = **6 dB** decrease

Worked example 8.5 illustrates the following general effect of attenuation from a point source of sound.

- **The SPL decreases by 6 dB each time the distance is doubled from a point source of sound in free space**

If the source is on a perfectly flat and reflecting surface then all the sound radiates into a hemisphere. The SPL then decreases by 3 dB (half of 6 dB) when the distance is doubled. A single motor car or other machine on a paved surface approximates to this condition and the sound attenuates by only 3 dB for each doubling of distance.

Line source of sound

The sound wave from a line source spreads out in the shape of a cylinder, as shown in figure 8.9. The surface area of a cylinder increases in simple proportion to its radius. Sound intensity from a line source therefore decreases in simple inverse proportion to the distance from the source.

The attenuation from a line source in a free field can be shown by general relationship

$$I \propto \frac{1}{d}$$

where I is the sound intensity measured at distance d from the source.

The ratio of any two intensities is given by the following formula:

$$\frac{I_1}{I_2} = \frac{d_2}{d_1}$$

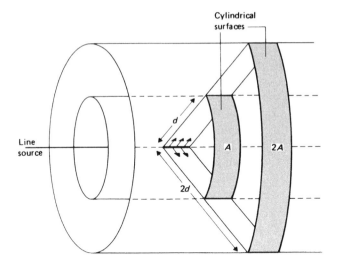

Surface area increases in proportion to the distance.
Intensity decreases in inverse proportion to the distance.

Figure 8.9 *Attenuation of sound from a line source*

where I_1 is the sound intensity measured at distance d_1 from the source and
I_2 is the sound intensity measured at distance d_2 from the source.

For a perfect line source the SPL decreases by 3 dB each time that the distance is doubled. For a line source radiating all its energy into half a cylinder the attenuation is therefore 1.5 dB (half of 3 dB) for each doubling of the distance. A line of cars on a busy road tends to approach this condition.

In practical situations near buildings, the decrease in sound level caused by doubling the distance is between 1.5 and 3 dB. Such changes in sound level of less than 3 dB are heard as a negligible change in loudness. Moving a building further away from the road therefore produces little effect within the area of most building plots and other methods, such as sound reflection and sound insulation, are more effective for the control of road noise.

Attenuation by air

The attenuation of sound caused by its spreading out from a point source or line source is an effect of energy distribution. The following factors may also affect the passage of sound through the air.

Air absorption
Some of the energy of a sound wave is spent in alternately compressing and expanding air. The effect is negligible at low frequencies and 2000 Hz

causes a reduction of about 0.01 dB/m of travel. The attenuation is increased at low humidities.

Temperature gradients

The velocity of sound is greater in warm air than in cold air. Open air has layers at different temperatures and sound waves crossing these layers are deflected by the process of *refraction.*

One result of this refraction is that, in general, sound travels along the ground better at night than during the day. This is the effect of relative changes in the temperature of the air lying next to the ground. Sound is refracted upwards during the day and at night it is refracted downwards.

Wind effects

Sound waves will be affected by any wind blowing between the source and the receiver. The velocity of wind increases with height above the ground and this gradient deflects the sound waves upwards or downwards.

Ground attenuation

It is possible for some sound energy to be absorbed by passing over the surface of the ground. This effect is quite local and only applies within 6 m of the ground, which must be free of obstructions. Hard surfaces, such as paving, provide little attenuation but surfaces such as grassland can provide a reduction of overall noise level of up to 5 dB.

NATURE OF HEARING

Sound waves are a phenomenon that can be detected and measured without the aid of human senses, but the aspect of sound that interests us most is the human perception of sound waves. The sense of hearing involves the ear and the brain, and the effect of sound can therefore vary from person to person. However, the basic characteristics outlined in this section are shared by most people.

The ear

Most of the mechanism of the ear is situated inside the head and the structure of the ear can be divided into three main parts, as shown diagrammatically in figure 8.10.

Outer ear

The outer ear is part of the ear that can be seen. It collects the sound waves and funnels them to the ear drum, a membrane which vibrates when sound waves fail upon it.

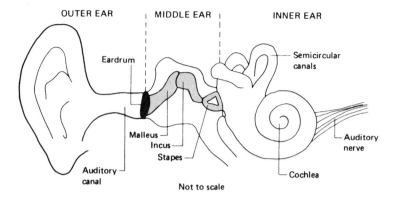

Figure 8.10 *The ear*

Middle ear

The middle ear is an air-filled cavity, connected to the throat, which passes the vibrations of the ear drum to the inner ear. This transfer is achieved by means of the three small bone levers. The mechanical link between the bones amplifies the vibrations to adjust for the difference between the air of the middle ear and the fluid of the inner ear.

Inner ear

The inner ear converts the mechanical vibrations of sound into electrical impulses which are transmitted to the brain. The *cochlea* is a hollow coil of bone, filled with liquid, in which the sound waves vibrate. Dividing the cochlea along its length is the *basilar membrane*, which contains approximately 25 000 nerve endings. The fine hairs attached to these nerves detect the sound vibrations in the fluid and the information is transmitted to the brain by the auditory nerve. The inner ear is situated near the three semi-circular canals which contain fluid and are associated with the sense of balance.

Deafness

Loss of hearing not only affects sensitivity to sounds but also affects the frequencies at which sounds can be detected. The various causes and categories of deafness are described below.

Middle ear deafness

Middle ear deafness is a result of a stiffening in the system of connecting bones. It may be caused by various infections or by a broken ear drum. The resulting deafness affects the transmission of low tones rather than high tones and can usually be cured by drugs or by surgery.

Nerve deafness

Nerve deafness is a result of damage to the nerve endings in the inner ear or to the nerve carrying information to the brain. This type of deafness can be caused by infections, by head injuries, and by exposure to high levels of noise.

Hearing loss caused by exposure to noise can give the following two effects:

- **TEMPORARY THRESHOLD SHIFT (TTS) is a temporary loss of hearing which recovers in 1–2 days after the exposure to noise**

- **PERMANENT THRESHOLD SHIFT (PTS) is a permanent loss of hearing caused by longer exposure to noise**

The first effect of *Noise Induced Hearing Loss* is loss of hearing in the region of high tones, around frequencies between 3000 and 4000 Hz. This effect may not affect speech reception at first and the deafness usually remains unnoticed until it begins to affect ordinary conversation. Because the receptor cells in the inner ear are damaged, this nerve deafness is irreversible and cannot be helped by amplification.

Presbyacusis

Presbyacusis is a gradual loss of hearing sensitivity due to age and is experienced by everybody. The higher frequencies are affected first but the effect is not usually noticeable until the age of 65 years or above. However, loss of hearing caused by age adds to any loss of hearing due to noise exposure earlier in life and the combination of the two may be noticeable.

Audiometry

Audiometry is the measurement of human hearing in terms of sensitivity at various frequencies across the noise spectrum. The threshold of hearing is measured at each frequency for each ear and the results can be presented graphically as an *audiogram.*

The results shown on an individual's audiogram can be compared with the average sensitivities for a person of the same age and with previous results for the same person. Such readings allow Noise Induced Hearing Loss caused by a person's job to be distinguished from the normal presbyacusis that occurs with age.

Loudness

As the intensity of a sound increases it is heard to be 'louder'. This sensation of loudness is a function of the ear and the brain and it depends upon the frequency as well as the amplitude of the sound wave. Human hearing is not equally sensitive at all frequencies and tones of different pitch will be judged to be of different loudness, even when their SPL is the same. For example, a 50 Hz tone must be boosted by 15 dB so as to sound equally loud as a 1000 Hz tone at 70 dB.

The results of many measurements of human hearing response can be presented in the form of standard contours, as in figure 8.11. The contours show how the SPL in dB of pure tones needs to change to create the same sensation of loudness when at different frequencies. It can be seen that the ear is most sensitive in the frequency range between 2 kHz and 5 kHz, and least sensitive at low frequencies or at extremely high frequencies. This effect is more pronounced at low SPLs than at high SPLs, and figure 8.11 has a 'family' of curves for different SPLs.

Phon scale

If two different tones seem equally loud to the ear then it can be useful to have a scale which gives them the same value, even though the two tones have different SPLs. The phon scale of loudness level is obtained from the family of equal loudness curves shown in figure 8.11.

- *LOUDNESS LEVEL (L_N)* **of any sound is numerically equal to that SPL, in decibels, of a 1000 Hz pure tone which an average listener judges to be equally loud**

  UNIT: phon

For example, it can be seen from figure 8.11 that

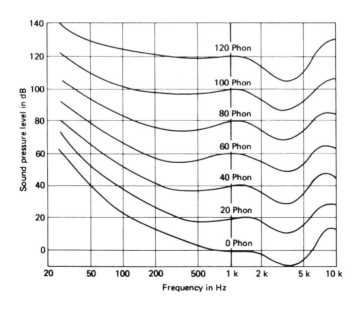

Figure 8.11 *Equal loudness contours*

60 phons = 60 dB at 1000 Hz

but

60 phons = 78 dB at 50 Hz

This difference between the phon and the decibel values reflects the fact that human hearing is less sensitive at low frequencies.

Sone scale

The *sone* scale of loudness is a re-numbering of the phon scale so that the sone values are directly proportional to the magnitude of the loudness. For example, a value of 2 sones is twice as loud as 1 sone. One sone is equivalent to 40 phons. Loudness, in sones, is doubled each time the loudness level is increased by 10 phons.

Exercises

1 A certain sound has an intensity of 3.16×10^{-4} W/m². Calculate the sound intensity level in decibels if the threshold of hearing reference intensity is 1×10^{-12} W/m².

2 Calculate the actual pressure of a sound which has an SPL of 72 dB. The threshold of hearing reference pressure is 20×10^{-6} Pa.

3 Calculate the total sound pressure level caused by the combination of sounds with the following SPLs:
   (a) 85 dB and 87 dB;
   (b) 90 dB and 90 dB and 90 dB.

4 The intensity of a point source of sound, measured at a distance of 6 m from the source, is 3.4×10^{-6} W/m².
   (a) Calculate the intensity of the sound at a distance 20 m from the source.
   (b) Calculate the decibel change between the two positions.

5 Calculate the difference between SPLs measured at 10 m and at 63 m from a perfect line source of sound. State the factors that might modify this result in a practical situation.

Answers

1 85 dB

2 0.0796 Pa

3 (a) 89.1 dB; (b) 94.7 dB

4 (a) 3.06×10^{-7} W/m²; (b) 10.5 dB

5 8 dB

9 Noise

Noise is unwanted sound. This is an environmental definition of sound that takes account of the effect of a sound rather than its nature. Even if a sound consists of the finest music it can be considered as noise if it occurs in the middle of the night!

Many of the reasons for not wanting a particular sound can be identified by the effects that it can have on the listener or on the environment. Some of these effects are described below:

- **Hearing loss:** Excessive exposure to noise causes loss of hearing, as discussed in chapter 8.
- **Quality of life:** Noisy environments, such as areas near busy roads or airports, are considered unpleasant and undesirable.
- **Interference:** Interference with significant sounds such as speech or music can be annoying and, in some situations, dangerous.
- **Distraction:** Distraction from a particular task can cause inefficiency and inattention which could be dangerous.
- **Expense:** The measures needed to control excessive noise are expensive. Businesses may also suffer loss of revenue in a noisy environment.

MEASUREMENT OF NOISE

The acceptance of noise by people obviously depends on individual hearing sensitivity and upon living habits. The acceptance of noise is also affected by the external factors outlined below:

- **Type of environment:** Acceptable levels of surrounding noise are affected by the type of activity. A library, for example, has different requirements from those for a workshop.
- **Frequency structure:** Different noises contain different frequencies and some frequencies are found to be more annoying or more harmful than others. For example, the high whining frequencies of certain machinery or jet engines are more annoying than lower frequency rumbles.

- **Duration:** A short period of high level noise is less likely to annoy than a long period. Such short exposure also causes less damage to hearing.

The measurement of noise must take these factors into account and the scale used to assess them should be appropriate to the type of situation. A number of different scales have evolved and this section describes the ones in common use.

Sound level meter

A sound level meter is an instrument designed to give constant and objective measurements of sound level. Figure 9.1 shows the main sections of a typical sound level meter. The meter converts the variations in air pressure to variations of voltage which are amplified and displayed on an electrical meter calibrated in decibels. Sound level meters can be small enough to be hand-held and are supplied in several grades of accuracy.

A typical sound level meter takes an RMS (root mean square) value of the signal, which is a type of average that is found to be more relevant than peak values. The meter may have 'fast' and 'slow' response settings but may not be fast enough to accurately record an impulse sound, such as gunfire.

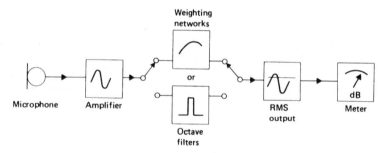

Figure 9.1 *Construction of a sound level meter*

Frequency components

Most practical noise is sound which contains a spectrum of different frequencies which are detected by the microphone of a sound level meter. The interpretation of these different frequencies needs consideration, especially because human hearing judges some frequencies to be more important than other frequencies.

If a noise contains *pure tones* of single frequencies then it is usually more annoying than broadband noise with a spectrum of frequencies. Pure tones are often present in the noise given off by industrial equipment such as high speed fans or other machinery, and by electrical generators and transformers.

Weighting networks

One method of dealing with the frequency content of sound is for the sound level meter to emphasise or give 'weight' to the same frequencies that human hearing emphasises. The weighting networks in a meter are electronic circuits whose response to frequency is similar to that of the ear. The response to low frequencies and to very high frequencies is reduced in a specified manner and four different weightings have been standardised internationally as the A, B, C and D scales.

The *A scale* has been found to be the most useful weighting network. Many measurements of noise incorporate decibels measured on the A scale and the symbols dB(A) indicate a standard treatment of frequency content.

Frequency bands

A sound level measurement that combines all frequencies into a single weighted reading is not suitable when it is necessary to measure particular frequencies. Some sound level meters allow the use of filters, which pass

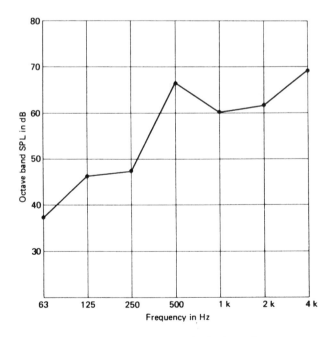

Figure 9.2 *Noise spectrum of telephone buzzer*

selected frequencies only. The sound pressure level (SPL) in decibels is then measured over a series of frequency bands, usually octaves or one-third octaves.

The *noise spectra* produced by such a series of measurements can be presented as a dB–frequency diagram, like that shown in figure 9.2. Narrower frequency bands give more information about the frequency content of the sound than wide frequency bands.

Noise limiting curves

One method of specifying acceptable levels of sound at different frequencies is to use standard curves of noise. These rating curves, which are plots of SPL against set frequencies, are based on the sensitivity of the human ear. The noise being analysed is measured at octave intervals and these results are plotted on top of the standard curves, as shown in figure 9.3.

The aim of most systems is to produce an index or criterion, which is a single-figure rating for the noise. This figure is obtained by comparing

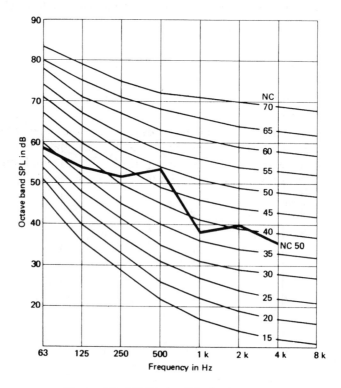

Figure 9.3 *NC (Noise Criterion) curves*

the plotted curve against the standard curves and determining the closest fit according to published rules. Commonly-used systems are as follows:

- **Noise Criterion (NC):** obtained from *NC curves*
- **Preferred Noise Criterion (PNC):** obtained from *PNC curves*
- **Noise Rating (NR):** obtained from *NR curves*

The NC curves have been widely used for assessing the noise made by heating and ventilating equipment, and PNC curves are a development of the NC curves. NR curves are commonly used for other industrial measurements of noise.

Acceptable noise limits for different purposes can be specified by numbers read from the standard curves and table 9.1 indicates some typical limits for levels of background noise caused by services installations.

Table 9.1 *Acceptable levels of background noise from services installations*

Environment	NC/NR/PNC approx. index
General offices	40
Libraries	35
Homes, hospitals	30
Theatres, cinemas	25
Concert halls, studios	20

Time components

Human response to noise greatly depends upon the total time of the noise and upon the variation in sound levels during that time. There is more tolerance, for example, of high but steady levels of background sound than of a lower level background level with frequent noise intrusions.

Statistical measurement

During a chosen period of time, such as 12 hours, it is possible to record many instantaneous readings of sound level, in dB(A). The variations in these readings can then be combined into one single number or 'index' by a statistical process using a *percentile level*, such as 10 per cent:

- **A PERCENTILE LEVEL, such as $L_{10,T}$, is the noise level exceeded for 10 per cent of a given measurement time T**

  UNIT: dB(A)

The traffic noise index, described below, uses the 10 per cent level for estimating maximum noise levels. Another percentile level, of 90 per cent, is used to estimate background noise.

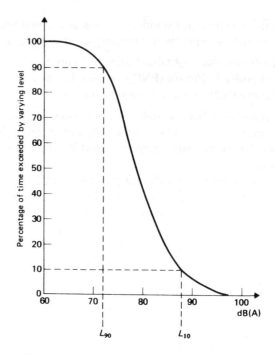

Figure 9.4 *Time distribution of noise level*

These statistical measurements are best made by a noise level analyser attached to the sound level meter which records the percentage of time the noise level has spent in each portion of the decibel scale. The analyser produces a single figure in dB(A) as the percentile level and some systems can also produce a graph of results, such as that shown in figure 9.4.

Traffic noise index L_{10}

Noise caused by road traffic usually changes in level during the day and the way it varies has a considerable effect on the nuisance it causes. The traffic noise index, L_{10}, takes these variations into account:

- **TRAFFIC NOISE INDEX is an average of the 18 hourly L_{10} values taken between 0600 and 2400 hours on a normal weekday**

Measurements of L_{10} must be taken at certain standard distances from the roadway with corrections made for any wind and reflections. A typical measurement at 10 m from the edge of a motorway is $L_{10,18hr}$ of 74 to 78 dB(A).

The traffic noise index has been found to give reasonable agreement with surveys of the dissatisfaction caused by road traffic noise. A typical basis for

Government compensation to households affected by new or improved roads is an $L_{10,18hr}$ of 68 dB(A).

Equivalent continuous sound level L_{eq}

Another method of assessing noise that varies in sound level over time is to use an average value related to total energy. The equivalent continuous sound level compares a varying sound level with a theoretical constant sound which gives an equivalent amount of sound energy.

Although human hearing does not judge loudness in terms of energy, the L_{eq} measurement of accumulated sound energy is found to correlate well to the annoyance caused by noise and also to hearing damage:

- **EQUIVALENT CONTINUOUS SOUND LEVEL, $L_{eq,T}$ is that constant sound level which, over the same period of time T, provides the same total sound energy as the varying sound being measured**

  UNIT: dB(A)

In a simple example, a doubling of sound energy increases the SPL by 3 dB, so that the following combinations of sound levels and time periods could give the same $L_{eq,T}$:

90 dB(A) for 8 hours
or
93 dB(A) for 4 hours   All give the same value of
or $L_{eq,8hr} = 90$ dB(A)
99 dB(A) for 1 hour

In many practical situations the noise levels tend to vary continuously and integrating sound level meters are employed to sample the noise regularly and to produce a L_{eq} figure as the 'noise history' over a particular period of time.

Hearing risk

Because the risk of damage to hearing is largely dependent on the total energy reaching the ear in a given period, L_{eq} is the basis of safe exposure to noise. Recommended levels are usually in the range L_{eq} (8 hour) = 80 to 90 dB(A), as described below.

Occupational noise index L$_{EP,d}$

The $L_{EP,d}$ index assesses the noise exposure of a worker in terms of the L_{eq} in dB(A) measured over 8 hours.

An $L_{EP,d}$ value of 85 db(A) is possibly hazardous and is used as an industrial *action level* when staff must be informed, assessments must be made, and hearing defenders made available.

Noise dose
The noise dose index makes 100 per cent dose equal to a fixed noise exposure such as 85 or 90db(A) for 8 hours. Individual noise exposure can be assessed by wearing a personal *dosemeter.*

Peak
The peak is the highest pressure produced by a an explosive sound, such as those from cartridge tools and gunshots. Although this peak only lasts for a brief instant, its pressure is high and damaging to hearing. The peak value is measured by a sound level meter which can hold and display the information.

A peak of 200 pascals (equivalent to 140 dB) is used as an industrial action level. At this level of sound, an 8 hour L_{eq} exposure of 90 db(A) is reached in about 1/5th second.

Construction site noise

A L_{eq} continuous equivalent sound level index is used to assess the noise associated with operations on a building site. The annoyance caused to the environment can be equated to the total sound energy received at the boundary of the site during the course of the day. A 12 hour L_{eq} value of 75 db(A) is a common limit, above which site operations can be stopped by legal action.

As with other cumulative measurements, different patterns of noise during the 12 hours can give the same value. For example, it is possible to run plant with sound levels over 100 dB(A) for several hours and still remain within the 12 hour L_{eq} of 75 dB(A). The prediction of such results can be made using published tables. Site measurements can be made at the edge of a site using the integrating sound level meters and techniques described in the sections above.

Other noise measurements

Sound Exposure Level (SEL)
The Sound Exposure Level is an index of transient noise levels, such as those produced by passing road vehicles or aircraft. The SEL is a measure of acoustic energy, which is a factor in the annoyance caused by a noise but is different from the maximum sound level in dB(A).

- **SOUND EXPOSURE LEVEL, L_{AE}, is that constant sound level in dB(A) which, during one second, provides the same total sound energy as the measured noise**

Perceived noise level (L_{PN})
L_{PN} is an index of aircraft noise that takes account of those higher frequencies in aircraft engine noise which are known to cause annoyance. The noisiness

is rated in *noys* and converted by calculation to perceived noise decibels PNdB.

- An approximate value for perceived noise level is obtained by adding 13 dB to the measured dB(A) noise level

Noise and Number Index (NNI)

NNI is an index of aircraft noise that includes the average perceived noise level and the *number* of flyovers heard in a given period. This index has been used to predict and to measure annoyance resulting from noise near airports. It has also been used as the basis for the award of compensation payments to householders for sound insulation.

Speech Interference Level (SIL)

SIL is a measure of the level of background noise at which the noise will interfere with speech in a particular situation. The type of voices and distances involved are taken into account.

NOISE CONTROL

Action against the many types of noise which cause concern in buildings can be grouped into three main areas:

- **Source:** Sources of sound may be outside the building, such as from a road, or may be within the building, such as noise from occupants.
- **Path:** The sound path may be through the air from the source to the building, or the path may be within the building.
- **Receiver:** The receiver of the sound may be the building itself, it may be a particular room, or it may be the person hearing the noise.

Sound reductions in all three areas of source, path and receiver are relevant to the design of quieter buildings, which is the concern of this chapter.

Noise levels in and around buildings are also affected by wider issues such as the design of quieter vehicles and machinery, the location of industry and transport, and the type of construction techniques in use. Sometimes these matters need to be enforced by regulations so that legislation also becomes a concern of noise control.

Noise control actions

The following actions are useful to control noise from many industrial sources, including factories and construction sites. The actions help prevent damage to the hearing of workers and to reduce noise pollution of the environment:

- **Elimination:** of noisy equipment and work methods
- **Substitution:** of quieter machinery and methods
- **Isolation:** of the plant by position, mountings and enclosures
- **Engineering:** modifications such as silencers and hush kits
- **Work practices:** such as short periods only of high noise levels
- **Personal protection:** such as hearing defenders.

NOISE TRANSFER

Noise is transferred into buildings and between different parts of buildings by means of several different mechanisms. It is necessary to identify the types of sound involved as being of two main types:

- Airborne sound, or
- Impact (structure-borne) sound.

Airborne sound

- *AIRBORNE SOUND* **is sound which travels through the air** *before* **reaching a partition**

Notice that this definition is not as simple as merely saying the 'sound travels through the air'. The vibrations in the partition under consideration must be started by sound which has travelled through the air. Sound transfer by airborne sound is shown in figure 9.5 and typical sources of airborne sound include voices, radios, musical instruments, traffic and aircraft noise.

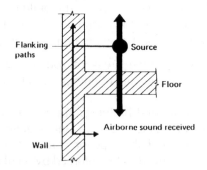

Figure 9.5 *Airborne sound transmission*

Impact sound

- *IMPACT SOUND* or *STRUCTURE-BORNE SOUND* **is sound which is generated** *on* **a partition**

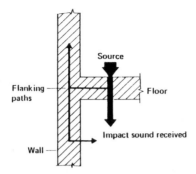

Figure 9.6 *Impact sound transmission*

The transfer of impact sound is shown in figure 9.6 and typical sources of impact sound include footsteps, slammed doors and windows, noisy pipes, and vibrating machinery. A continuous vibration can be considered as a series of impacts and impact sound is also termed structure-borne sound. Notice that such impact sound will normally travel through the air to reach your ear; but it is not the same as airborne sound.

It is important to distinguish between airborne sound and impact sound because the best methods of controlling them can differ. A single source of noise may also generate both types of sound so the definition of airborne and impact sound must be applied to the sound which is being heard in the *receiving* room. For example, footsteps on a floor would be heard mainly as impact sound in the room below but heard as airborne sound in the room above.

Sound can also pass into a receiving room by *flanking transmission*, as shown in figures 9.5 and 9.6. These indirect sound paths can be numerous and complex. The effect of flanking transmissions increases at high levels of sound insulation and often limits the overall noise reduction that is possible.

Absorption and insulation

The techniques used to control sound are described by some terms that may appear to be interchangeable but are, in fact, very different in their effect. Poor understanding of these terms leads to incorrect and wasted efforts in the control of sound.

Sound insulation

- **SOUND INSULATION is the reduction of sound energy transmitted into an adjoining air space**

Insulation is the most useful method for controlling noise in buildings and is discussed in the next section.

Sound absorption

- **SOUND ABSORPTION is a reduction in the sound energy reflected by the surfaces of a room**

Absorption has little effect on noise control but has an important effect on sound quality, which is discussed in chapter 10 on acoustics.

It can be shown by calculation that if the amount of absorption in a room is doubled then the sound energy in the room is halved, but the sound level drops by only 3 dB. Such a change in absorption may, however, make a difference to the subjective or apparent sound level in the room because this is influenced by the acoustic quality.

Sound absorption can also be useful in the control of noise, which spreads by reflections from the ceiling of offices and factories, or by multiple reflections along corridors or ducts, as shown in figure 9.7.

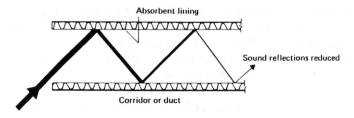

Figure 9.7 *Noise control by absorption*

SOUND INSULATION

Insulation is the principal method of controlling both airborne and impact sound in buildings. The overall sound insulation of a structure depends upon its performance in reducing the airborne and impact sound transferred by all sound paths, direct and indirect. The assessment of sound insulation initially considers one type of sound transfer at a time.

Sound reduction index

The difference in sound levels on either side of a partition, as shown in figure 9.8, can be used as an index of airborne sound insulation.

- **SOUND REDUCTION INDEX (SRI) is a measure of the insulation against the direct transmission of airborne sound**

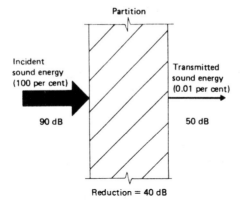

Figure 9.8 *Airborne sound insulation*

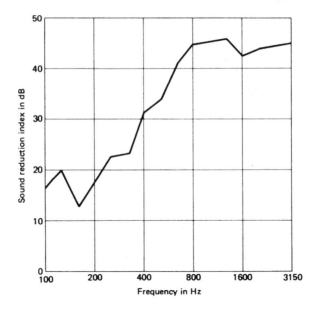

Figure 9.9 *An insulation curve for double glazing*

The measurement of an SRI can be made in a special laboratory where no flanking sound paths are possible around the partition under test. The sound levels in the rooms on each side of the partition are measured and a *normalised* SRI obtained by adjusting for the area of the partition and for the absorption in the receiving room.

Table 9.2 *Sound insulation values*

Construction	Surface mass nominal (kg/m^2)	SRI 100–3010 Hz average (dB)
Walls		
255 mm brick/cavity/brick, plastered, wire ties	425	53
215 mm brickwork, plastered	425	50
102.5 mm brickwork, plastered	215	45
100 mm dense concrete	230	45
300 mm lightweight concrete	190	42
50 mm dense concrete	115	40
12 mm plasterboard and plaster	12	25
Windows		
Double, 150–200 mm air gap, sealed		40 max.
Single, 12 mm glass, sealed	30	30
Single, 6 mm glass, sealed	15	25
Single, unsealed		20
Any window, open		10 approx.

Note
Sealed means fixed; or openable but weather-stripped.

Because insulation varies with frequency, the SRI is measured at octave intervals between 100 Hz and 3150 Hz and then plotted against frequency, as shown in figure 9.9. The arithmetic average of these SRIs is usually similar to the single value measured at 500 Hz and this average value is often convenient for initial calculations. Some typical average SRIs are quoted in table 9.2 and the calculation of such insulation values is treated later in the chapter.

Insulation principles

Good sound insulation depends upon the following general principles:

- Heaviness
- Completeness
- Flexibility
- Isolation.

The effectiveness of each technique of insulation can differ with the type of sound but in most constructions all the principles are relevant. The principles are described in the following sections.

Heaviness

Heavyweight structures with high mass transmit less sound energy than lightweight structures. The high density of heavyweight materials restricts the size of the sound vibrations inside the material so that the final face of the structure, such as the inside wall of a room, vibrates with less movement than for a lightweight material.

Because the vibrations of this 'loudspeaker' effect are restricted, the amplitude of the sound waves re-radiated into the air is also restricted. Although a reduction in the amplitude of sound waves affects the 'strength' or 'loudness' of a sound, it does not affect the frequency (pitch) of that sound.

Mass law
- *THE MASS LAW*: **The sound insulation of a single leaf partition is proportional to its mass per unit area**

Single-leaf construction includes composite construction such as plastered brickwork, as long as the layers are bonded together. Theory predicts an insulation increase of 6 dB for each doubling of mass but for practical constructions the following working rule is more suitable:

- Sound insulation increases by 5 dB whenever the mass is doubled.

For example, the average SRI of a brick wall increases from 45 dB to 50 dB when the thickness is increased from 102.5 mm to 215 mm. This doubling of mass does not have to be achieved by a doubling of thickness as the mass of a wall for sound insulation purposes is specified by its *surface density* measured in kilograms per square metre (rather than per cubic metre). Concrete blocks of different densities can produce the same surface density by varying the thicknesses of the blocks.

The effectiveness of sound insulation depends upon frequency and the Mass Law also predicts the following effect on frequency:

- Sound insulation increases by about 5 dB whenever the frequency is doubled.

Any doubling of frequency is a change of one *octave*. For example, a brick wall provides about 10 dB more insulation against 400 Hz sounds than against 100 Hz sounds. This change, a rise of two octaves.

Table 9.2 gives the surface masses, or construction densities, of typical structures. Where a construction does not obey the Mass Law it is because other factors such as airtightness, stiffness, and isolation have an effect.

Completeness

Areas of reduced insulation or small gaps in the construction of a wall have a far greater effect on overall insulation than is usually appreciated. The completeness of a structure depends upon airtightness and uniformity.

Airtightness

As insulation against airborne sound is increased the presence of gaps becomes more significant. For example, if a brick wall contains a hole or crack which in size represents only 0.1 per cent of the total area of the wall, the average SRI of that wall is reduced from 50 dB to 30 dB.

Air gaps often exist because of poorly constructed seals around partitions, particularly at the joins with floors, ceilings, windows, doors, service pipes, and ducts. Some materials may be also porous enough to pass sound through the small holes in their structure; brick and blockwork should therefore be plastered or sealed. Doors and openable windows should be airtight when closed and the type of sealing used to increase thermal insulation is also effective for sound insulation. In general, 'sound leaks' should be considered as carefully as leaks of water.

Uniformity

The overall sound insulation of a construction is greatly reduced by small areas of poor insulation. For example, an unsealed door occupying 25 per cent of the area of a half-brick wall reduces the average SRI of that wall from around 45 dB to 23 dB. The final sound insulation is influenced by relative areas but is always closer to the insulation of the poorer component than to the better component.

Windows and doors are necessary parts of a building but a knowledge of the uniformity principle can prevent effort being wasted on the insulation of the wrong areas. To improve the insulation of a composite structure the component with the lowest insulation should be improved first of all. Walls facing noisy roads should contain the minimum of windows and doors, and they should be well insulated.

Flexibility

Stiffness is a physical property of a partition and depends upon factors such as the elasticity of the materials and the fixing of the partition. High stiffness can cause loss of insulation at certain frequencies where there are resonances and coincidence effects. These effects upset the predictions of the Mass Law, as indicated in figure 9.10.

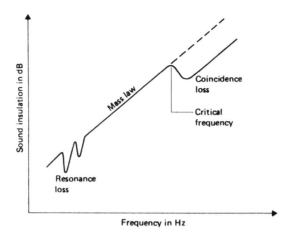

Figure 9.10 *Resonance and coincidence loss*

Resonance

Loss of insulation by resonance occurs if the incident sound waves have the same frequency as the natural frequency of the partition. The increased vibrations that occur in the structure are passed on to the air and so the insulation is lowered. Resonant frequencies are usually low and most likely to cause trouble in the air spaces of cavity construction.

Coincidence

Loss of insulation by coincidence is caused by the bending flexural vibrations, which can occur along the length of a partition. When sound waves reach a partition at angles other than 90 degrees, their transmission can be amplified by the flexing inwards and outwards of the partitions. The sound-wave frequency and the bending-wave frequency coincide at the *critical frequency*. For several octaves above this critical frequency the sound insulation tends to remain constant and less than that predicted by the Mass Law. Coincidence loss is greatest in double-leaf constructions, such as in cavity walls or in hollow blocks.

Flexible (non-stiff) materials, combined with a high mass, are best for high sound insulation. Unfortunately flexibility is not usually a desirable structural property in a wall or a floor.

Isolation

Discontinuous construction can be effective in reducing the transmission of sound through a structure. As the sound is converted to different wave motions at the junction of different materials energy is lost and a useful amount of insulation is gained. This is the principle behind the effectiveness

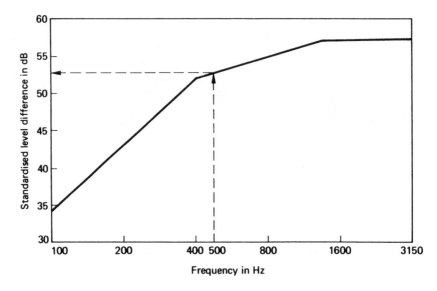

Figure 9.11 *Reference curve for airborne sound insulation*

of air cavities in windows, of floating floors, of carpets, and of resilient mountings for vibrating machines. Some broadcasting and concert buildings achieve very high insulation by using the completely discontinuous construction of a double structure separated by resilient mountings.

Sound isolation is easily ruined by strong *flanking* transmissions through rigid links, even by a single nail. Cavity constructions must be sufficiently wide for the air to be flexible, otherwise resonance and coincidence effects can cause the insulation to be reduced at certain frequencies.

Sound insulation regulations

Building Regulations, such as those for England and Wales, require certain minimum standards of sound insulation for those walls and floors which protect the rooms in dwellings where people live. The distinction between airborne and impact sound, as explained earlier in the chapter, is important and figure 9.11 shows the major requirements.

The requirements of sound-insulation regulations may be satisfied by adopting one of the widely-used constructions, some of which are listed in table 9.3. Attention must be paid to details of the construction, especially at junctions, so as to avoid flanking transmission of sound.

Another method of satisfying the regulations is to repeat a form of construction which already has been used in a similar building and shown, by

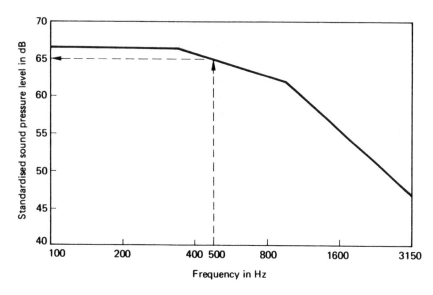

Figure 9.12 *Reference curve for impact sound insulation*

Table 9.3 *Some construction suitable for sound insulation*

Construction	Min. density of masonry
Walls	
Solid brickwork, pbs	375 kg/m^2
Concrete blockwork, pbs	415 kg/m^2
Brick cavity, pbs	415 kg/m^2
Lightweight concrete blockwork cavity, pbs	415 kg/m^2
Lightweight concrete blockwork, with lightweight panels each side	200 kg/m^2
Timber frame, mineral fibre quilt inside, with double-thickness plasterboard each side	N/A
Floors	
Concrete base, soft covering	365 kg/m^2
Concrete base with floating screed	220 kg/m^2
Concrete base with floating timber raft	220 kg/m^2
Timber base, absorbent layer with floating layer	N/A

Note
pbs = plastered both sides.

Table 9.4 *Required sound transmission values*

Type of transmission (4 pairs of rooms)	Individual values	Mean values
AIRBORNE SOUND (minimum values)	49 (walls) 48 (floors)	53 (walls) 52 (floors)
IMPACT SOUND (maximum values)	61	65

Notes
Airborne sound values are weighted standardised level difference (D_{nTw}).
Impact sound values are weighted standardised sound pressure
level (L_{nTw}).

tests, to meet the requirements listed in table 9.4. The insulation of the construction is measured on site by established procedures such as those described in British Standards.

For airborne sound, the results are given as the difference in sound level between the room with the sound source and the receiving room. For impact sound, a measurement is made of the sound received from a standard source of impact sound. Results are adjusted to take account of the acoustic effect of the receiving room and this adjusted figure is known as the *standardised* or *normalised* reading.

Measurements are made at 16 frequencies between 100 and 3150 kHz. The measurements are plotted and compared with a standard reference curve in accordance with certain rules so that a single figure at 500 Hz can be quoted. These single figures, quoted in table 9.4, are known as the *weighted* standardised measurements.

Insulated constructions

Figures 9.13, 9.14, 9.15 and 9.16 show the details of some forms of construction that are considered to have adequate performance for sound insulation. The function of the special construction features should be evaluated in terms of the principles of sound insulation described in the previous sections.

Walls
Heavyweight walls generally provide high levels of airborne sound insulation because of the effect of the Mass Law. Cavities can increase sound insulation by the principle of isolation but practical cavity walls with wire ties behave little better in practice than a double skin of bricks.

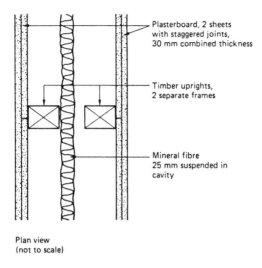

Plasterboard, 2 sheets
with staggered joints,
30 mm combined thickness

Timber uprights,
2 separate frames

Mineral fibre
25 mm suspended in
cavity

Plan view
(not to scale)

Figure 9.13 *Insulated timber frame wall*

Lightweight walls can provide adequate levels of airborne sound insulation if attention is paid to the principles of isolation and airtightness. Each side of the wall is built on separate timber frames and an absorbent blanket of mineral fibre is used to provide acoustical isolation between the sides of the wall. The mass provided by multiple layers of dense plasterboard contributes significantly to the insulation of the wall.

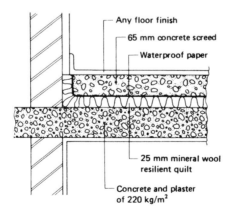

Any floor finish

65 mm concrete screed

Waterproof paper

25 mm mineral wool
resilient quilt

Concrete and plaster
of 220 kg/m^2

Figure 9.14 *Insulated concrete floor*

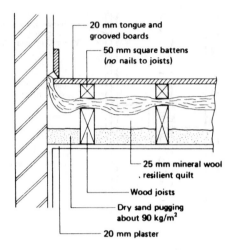

Figure 9.15 *Insulated wood-joist floor*

Floors

The natural mass of concrete floors provides airborne sound insulation and a resilient layer is also needed to provide insulation against impact sound. The mass of wooden floors can be increased by the use of sand *pugging* within the joist space. More recent methods of mass enhancement include the use of thick layers of composite floorboard or multiple layers of plasterboard.

The insulation of a floor must be maintained at all junctions with the surrounding walls in order to prevent flanking transmission. The separation of the two parts of a floating floor must continue around all the edges by the use of resilient materials and airtight techniques.

Windows

Air cavities must be greater than 150mm wide to provide worthwhile isolation and the cavity should be lined with absorbent material to minimise resonance. Heavyweight glass provides increased sound insulation because of the Mass Law. The frames must be isolated from one another by some construction technique such as a resilient layer while good fittings and seals provide airtightness.

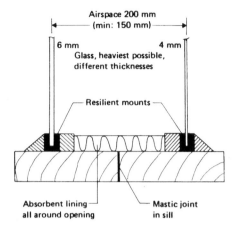

Figure 9.16 *Insulated double window*

Calculation of sound insulation

The sound reduction index used to measure the airborne sound insulation of a partition depends upon the amount of sound energy transmitted across the partition, as shown in figure 9.8.

The proportion of energy transmitted through the partition is measured by the transmission coefficient, *T*, where

$$T = \frac{\text{transmitted sound energy}}{\text{incident sound energy}}$$

The *sound reduction index* is then defined by the following formula:

$$\text{SRI} = 10 \log_{10}\left(\frac{1}{T}\right)$$

UNIT: decibel (dB)

Worked example 9.1

A wall transmits 1 per cent of the sound energy incident upon the wall at a given frequency. Calculate the sound reduction index of the wall at this frequency.

Let   Incident sound energy = 100
         Transmitted sound energy = 1

So *T* = (transmitted sound/incident sound) = 1/100 = 0.01

Using

$$SRI = 10 \log(1/T)$$

$$SRI = 10 \log(1/0.01) = 10 \log(100)$$

$$= 10 \times 2$$

So SRI = **20 dB**

Composite partitions

A window placed in a well-insulated wall can greatly reduce the overall sound insulation of the wall, as was discussed in the section on completeness. Because SRIs have been calculated using a logarithmic formula they cannot be simply averaged according to area. But the overall transmission coefficient can be calculated using the transmission coefficients and areas of the individual components, such as shown in figure 9.17.

$$T_0 = \frac{(T_1 \times A_1) + (T_2 \times A_2) + (T_3 \times A_3)}{A_1 + A_2 + A_3}$$

where T_0 = overall transmission coefficient
T_1 = transmission coefficient of one component
A_1 = area of that component etc.

The overall sound reduction index for the complete partition can then be calculated using the overall transmission coefficient.

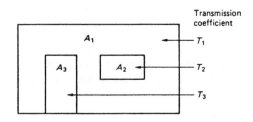

Figure. 9.17 *Sound insulation of composite partition*

Worked example 9.2

A wall of area $10\,m^2$ contains a window of area $2\,m^2$. The SRIs are: $50\,dB$ for the brickwork and $18\,dB$ for the window. Calculate the overall SRI for the wall.

For brickwork let: $\quad T_1 = ?, \qquad A_1 = 10 - 2 = 8\,m^2, \qquad SRI = 50\,dB.$
For window let: $\qquad T_2 = ?, \qquad A_2 = 2\,m^2, \qquad\qquad SRI = 18\,dB.$

Using $SRI = 10\log(1/T)$

$$50 = 10\log(1/T_1) \qquad \text{and} \qquad 18 = 10\log(1/T_2)$$

$$\log\frac{1}{T_1} = 5 \qquad\qquad\qquad \log\frac{1}{T_2} = 1.8$$

$$\frac{1}{T_1} = 10^5 \qquad\qquad\qquad \frac{1}{T_2} = 63.10$$

$$T_1 = 10^{-5} \qquad\qquad\qquad T_2 = 1.585 \times 10^{-2}$$

Using

$$T_0 = \frac{(T_1 \times A_1) + (T_2 \times A_2)}{A_1 + A_2}$$

$$= \frac{(10^{-5} \times 8) + (1.585 \times 10^{-2} \times 2)}{8 + 2} = \frac{3.18 \times 10^{-2}}{10}$$

$$= 3.18 \times 10^{-3}$$

Using

$$SRI = 10\log\left(1/T_0\right)$$

$$= 10\log\frac{1}{3.18 \times 10^{-3}} = 10\log 314.5 = 25$$

So the overall SRI = **25 dB**

Exercises

1 Use the information given in table 9.2 to plot points on a graph of sound insulation index against surface mass. Draw a best-fit line through the points to show the Mass Law relationship. Explain why some of the structures do not follow the Mass Law.

2 When the SRIs are measured for a certain double glazing unit the results obtained are those shown in figure 9.9. Explain the reason for the dips in

the insulation curve and outline some techniques that would help to reduce these effects.

3 Use the information given in table 9.3 to draw an annotated section of any wall that is considered to give sufficient sound insulation between dwellings. Explain which features of this wall help to provide the insulation.

4 800 units of sound energy are incident upon a wall and 10 of these units are transmitted through the wall.
   (a) Calculate the SRI of this wall.
   (b) If a window has a SRI of 33 dB then calculate the transmission coefficient of this window.

5 An external brick cavity wall is to be 4 m long and 2.5 m high. The wall is to contain one window 1.2 m by 800 mm and one door 750 mm by 2 m. The relevant sound reduction indexes are: brickwork 53 dB; window 25 dB; door 20 dB. Calculate the overall SRI of the completed partition.

Answers

4 (a) 19 dB; (b) 5.012×10^{-4}

5 27.4 dB

10 Room Acoustics

The term 'acoustics' can be used to describe the study of sound in general but the subject of room acoustics is concerned with the control of sound *within* an enclosed space. The general aim is to provide the best conditions for the production and the reception of desirable sounds. Noise control was treated separately in chapter 9 but the exclusion of unwanted noise is an important element of room acoustics. Similarly, the acoustic quality of sound in a room can affect the way that people judge noise levels.

The sound quality of a large auditorium, such as a concert hall, can be difficult to perfect and acoustics has sometimes been described as an 'art' rather than a 'science'. But, as with thermal and visual comfort, there are technical properties which do affect our perception and these make the best starting point for designing or improving the environment.

ACOUSTIC PRINCIPLES

General requirements

The detailed acoustic requirements for a particular room depend upon the nature and the purpose of the space, and the exact nature of a 'good' sound is partly a matter of personal preference. The general requirements for good acoustics are summarised as follows:

- Adequate levels of sound
- Even distribution to all listeners in the room
- Rate of decay (reverberation time) suitable for the type of room
- Background noise and external noise reduced to acceptable levels
- Absence of echoes and similar acoustic defects.

Types of auditorium

An auditorium is a room, usually large, designed to be occupied by an audience. The acoustic design of auditoria is particularly important and detailed acoustic requirements vary with the purpose of the space, as outlined below.

221

Speech

The overall requirement for the good reception of speech is that the speech is intelligible. This quality will depend upon the power and the clarity of the sounds. Examples of auditoria that are especially used for speech are conference halls, law courts, theatres, and lecture rooms.

Music

There are many more acoustic requirements for music than for speech. Music consists of a wide range of sound levels and frequencies which all need to be heard. In addition, some desirable qualities of music depend upon the individual listener's judgement and taste. These qualities are difficult to define but terms in common use include 'fullness' of tone, 'definition' of sounds, 'blend' of sounds, and 'balance' of sounds.

Examples of auditoria designed exclusively for music are concert halls, opera houses, recording studios, and practice rooms.

Multi-purpose auditoria

The previous sections indicate that there are some conflicts between the ideal acoustic conditions for music and for speech. Compromises have to be made in the design of auditoria for more than one purpose and the relative importance of each activity needs to be decided upon. Churches, town halls, conference centres, school halls, and some theatres are examples of multi-purpose auditoria.

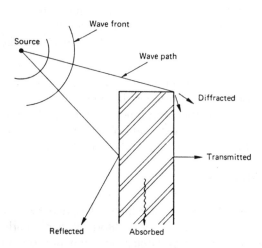

Figure. 10.1 *Sound paths*

Sound paths in rooms

A *sound path* or *sound ray* is the directional track made by the wave vibrations as they travel through a material such as air. The scaled geometrical drawing of sound rays is a useful technique for predicting acoustic effects. The behaviour of sound paths inside an enclosed space can be affected by the mechanisms of reflection, absorption, transmission, and diffraction, as shown in figure 10.1. As with other wave forms, such as light, *diffraction* is an effect which occurs at the edges of objects and is one reason why it is possible to hear sounds around corners.

Reflection and *absorption* play the largest roles in room acoustics, the final result depending upon the particular size and shape of the enclosure, and the nature of the materials used for the surfaces.

REFLECTION

Sound is reflected in the same way as light, provided that the reflecting object is larger than the wavelength of the sound concerned. The angle of reflection equals the angle of incidence of the wave, as shown in figure 5.10 for light rays. Remember that this angle is measured from a line drawn at right angles to the surface (the normal) and not measured from the surface itself. Using the rule of reflection, the straight lines representing sound rays can be drawn on plans and used to predict some of the effects of reflection. The special case of rapid reflections or reverberation is treated in a later section.

Types of reflector

Reflecting surfaces in a room are used to help the even distribution of sound and to increase the overall sound levels by reinforcement of the sound waves. The following sections describe various reflection effects and also indicate that there can be unwanted reflections (echoes). The following general rules apply:

- Reflections near the source of sound can be useful
- Reflections at a distance from the source may be troublesome.

Plane reflectors

The effect of a plane or flat reflector is shown in figure 10.2. Adjustable plane reflectors are often suspended above a stage to give an early reflection into the audience. Reflectors should be wide enough to reflect sound across the full width of the audience and the reflected sound must not be significantly delayed.

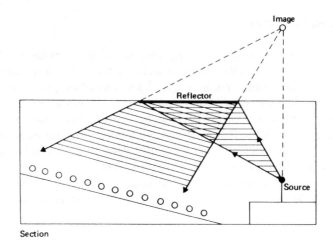

Figure 10.2 *Plane reflector of sound*

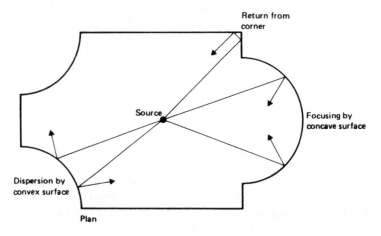

Figure 10.3 *Reflection from room surfaces*

Curved reflectors

At curved surfaces the angle of reflection still equals the angle of incidence but the geometry gives varying effects. Figure 10.3 illustrates the main type of reflection and shows the two types of curved surface.

- **Concave surfaces tend to focus sound**
- **Convex surfaces tend to disperse sound.**

Convex surfaces (outwards curving) are useful for distributing sound over larger areas. When the curves are of tight radius the sound will be scattered at random, which is a useful effect of the ornaments and elaborate plasterwork found in early theatres and cinemas.

Concave surfaces (inwards curving) tend to concentrate sound energy in particular areas and this is often acoustically risky. The domed ceilings of some public buildings, such as the Royal Albert Hall in London, have often contributed to unsatisfactory acoustics and required remedies. A shallow concave surface may be satisfactory if the focal point is well outside the enclosure and the tight curves found in decoration will contribute to scattering the sound.

Echoes

An echo is a delayed reflection. Initially a reflected sound reinforces the direct sound but if the reflection is delayed and is strong then this echo causes blurring and confusion of the original sound. The perception of echoes depends upon the power and the frequency of sounds.

There is a risk of a distinct echo if a strong reflection is received later than 1/20th second (50 ms) after the reception of the direct sound. At a velocity of 340 m/s this time difference corresponds to a path difference of 17 m.

This difference in length between direct sound paths and reflected paths can be checked by geometry and is most likely to affect seats near the front of a large auditorium. Late reflections can be minimised by the use of absorbers on those surfaces that cause the echoes.

Flutter echoes are rapid reflections which cause a 'buzzing' quality as sound decays. They are caused by repeated reflections between smooth parallel surfaces, especially in smaller rooms. The flutter can be avoided by using dispersion and absorption at the surfaces.

Standing waves

Each frequency of a sound has a wavelength. Sometimes the distance between parallel walls in a room may equal the length of half a wave, or a multiple of a half wavelength. Repeated reflections between the surfaces cause standing waves or *room resonances*, which are detected as large variations in sound level at different positions. Standing wave effects are most noticeable for low-frequency sounds in smaller rooms and, in general, parallel reflecting surfaces should be avoided.

Hall shapes

Acoustic requirements are not the only factors deciding the internal shape of an auditorium. Everyone needs to see the stage, for example, so good sightlines from seats are a major requirement. Fortunately, this requirement of being able to see the stage also helps to provide a strong component of *direct sound* which is an important feature of good acoustics. The aspects of shape described below are likely to affect the acoustics of an auditorium. Figure 10.4 shows one possible plan and section for a concert hall which includes many of the features necessary for good acoustics.

Rectangular

A rectangular plan or 'shoebox' is the traditional shape of many successful older concert halls. The sound waves tend to establish themselves along the length of the hall and all listeners receive a strong component of direct sound. Reflectors can be used to direct sound to the rear of the hall and absorbers used to prevent unwanted reflections. Traditional ratios of dimensions for height, width and length are about 2:3:5.

Wide fan

A fan shape or short-wide hall allows more people in the audience to be near the source of sound and gives them better views of the stage. Wide curves were used for the seating of ancient outdoor theatres and amphithea-

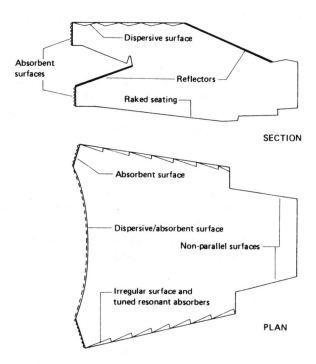

Figure 10.4 *Acoustic features of a concert hall*

tres and can give agreeable acoustics, although the acoustics are not always optimised for music.

Horse shoe

A horse-shoe shape is common in traditional opera houses where the tall concave shape at the back of the hall is broken up by tiers of seats and boxes. The audience and furnishings in these tiers also act as an absorber.

Raked seats

Most theatres have their rows of seats rising towards the rear. This raking provides a good view of the stage and prevents absorption of direct sound paths. Raked seating is not essential for music quality and the floor of a traditional opera house is relatively flat.

ABSORPTION

Sound absorption is a reduction in the sound energy reflected from a surface. In chapter 9 sound absorption was distinguished from sound insulation because the two concepts have different effects and applications. Sound absorption is a major factor in producing good room acoustics, especially when controlling reverberation.

Absorption coefficient

The absorption coefficient is a measure of the amount of sound absorption provided by a particular type of surface. The amount of sound energy not reflected is compared with the amount of sound energy arriving at the surface in the following formula:

Absorption coefficient $(a) = \dfrac{\text{Absorbed sound energy}}{\text{Incident sound energy}}$

UNIT: none – its value is expressed as a ratio

- $a = 1$ is the maximum for a perfect absorber
- $a = 0$ is the minimum for a poor absorber (or perfect reflector)

A surface that absorbs 40 per cent of incident sound energy has an absorption coefficient of 0.4. Note that the coefficient of 'absorption' is a surface consideration and is not affected by what actually happens to the sound energy that is not reflected. A strange example of a perfect absorber, for example, is an open window.

Different materials and constructions have different absorption coefficients, and the coefficient for any one material varies with the frequency of the incident sound. Table 10.1 lists the average absorption coefficients of

Table 10.1 *Absorption coefficients*

Common building materials		Absorption coefficient		
		125 Hz	500 Hz	2000 Hz
Brickwork	plain	0.02	0.03	0.04
Clinker blocks	plain	0.02	0.06	0.05
Concrete	plain	0.02	0.02	0.05
Cork	tiles 19 mm, solid backing	0.02	0.05	0.10
Carpet	thick pile	0.10	0.50	0.60
Curtains	medium weight, folded	0.10	0.40	0.50
	medium weight, straight	0.05	0.10	0.20
Fibreboard	13 mm, solid backing	0.05	0.15	0.30
	13 mm, 25 mm airspace	0.30	0.35	0.30
Glass	4 mm, in window tiles,	0.30	0.10	0.07
	solid backing	0.01	0.01	0.02
Glass fibre	25 mm slab	0.10	0.50	0.70
Hardboard	on battens, 25 mm airspace	0.20	0.15	0.10
Plaster	lime or plaster, solid backing	0.02	0.02	0.04
	on laths/studs, airspace	0.30	0.10	0.04
Plaster tiles	unperforated, airspace	0.45	0.80	0.65
Polystyrene tiles	unperforated, airspace	0.05	0.40	0.20
Water	swimming pool	0.01	0.01	0.01
Wood blocks	solid floor	0.02	0.05	0.10
Wood boards	on joists/battens	0.15	0.10	0.10
Wood wool	25 mm slab, solid backing	0.10	0.40	0.60
	25 mm slab, airspace	0.10	0.60	0.60
Special items				
Air	per m^3			0.007
Audience	per person	0.21	0.46	0.51
Seats	empty fabric, per seat	0.12	0.28	0.28
	empty metal, canvas, per seat	0.07	0.15	0.18

some common materials at the standard frequencies used in acoustic studies.

Total absorption

The effective absorption of a particular surface depends on the both the absorption coefficient of the surface material and the area of that particular surface exposed to the sound. A measure of this total absorption is obtained by multiplying the two factors together.

Absorption of surface = Area of surface × Absorption coefficient of
that surface

UNITS: m^2 sabins or 'absorption units'

The total absorption of a room is the sum of the absorptions provided by each surface in the room. This total is the sum of the products of all areas and their respective absorption coefficients as expressed in the following formula:

- Total absorption = Σ (Area × Absorption coefficient)

This calculation is usually best set out in a table, as shown later in the chapter, or on a computer spreadsheet. People and soft furnishings absorb sound, and air also absorbs sound at higher frequencies. Absorption factors for these items are given in table 10.1.

Types of absorber

The materials and the devices used especially for the purpose of absorbing sound can be classified into the following three main types of absorber which have maximum effect at different frequencies, as indicated in figure 10.5:

- Porous absorbers for high frequencies
- Panel absorbers for lower frequencies
- Cavity absorbers for specific lower frequencies.

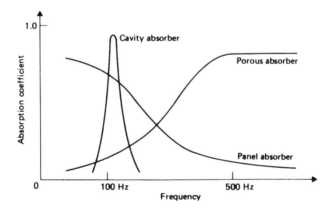

Figure. 10.5 *Response of different absorbers*

Porous absorbers

Porous absorbers consist of cellular materials such as fibreglass and mineral wool. The air in the cells provides a viscous resistance to the sound waves which then lose energy as frictional heat. The cells should interconnect with one another and the closed cell structure of some foamed plastics is not always the most effective form for sound absorption.

Porous materials used for sound absorption include acoustic tiles, acoustic blankets, and special coatings such as acoustic plaster. The absorption of porous materials is most effective at frequencies above 1 kHz; the low frequency absorption can be improved slightly by using increased thickness of material.

Panel absorbers

Panel or *membrane* absorbers are constructed from fixed sheets of continuous materials with a space behind them; the space may be of air or may contain porous absorbent. The panels absorb the energy of sound waves by converting them to mechanical vibrations in the panel which, in turn, lose their energy as friction in the clamping system of the panel. The panels may be made of materials such as plywood or they may already exist, as for example in the form of windows or suspended ceilings.

The amount of absorption depends on the degree of damping in the system. The *resonant frequency* of the system, at which maximum absorption occurs, depends on the mass of the panel and the depth of the airspace beyond. The resonant frequency is given by the formula

$$f = \frac{60}{\sqrt{(md)}}$$

where m and d are the measurements shown in figure 10.6.

A panel absorber is most effective for low frequencies in the range 40 to 400 Hz. A typical response curve is shown in figure 10.5.

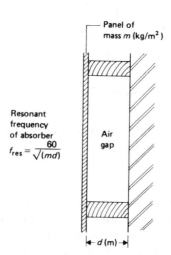

Figure 10.6 *Panel absorber*

Cavity absorbers

Cavity absorbers or *Helmholtz resonators* are enclosures of air with one narrow opening. The opening acts as an absorber when air in the opening is forced to vibrate and the viscous drag of the air removes energy from the sound waves. In practice, the cavity may contain material other than air and be part of a continuous structure, such as in a perforated acoustic tile.

A cavity can provide a high absorption coefficient over a very narrow band of frequencies. The maximum absorption occurs at the resonant frequency of the cavity which is estimated by the formula

$$f = 55\sqrt{(a/dV)}$$

where the measurements are those shown in figure 10.7. The ability to tune a cavity absorber to specific frequencies is useful for controlling certain sounds inside rooms; a typical response curve is shown in figure 10.5.

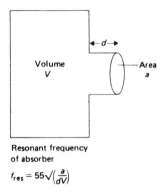

Resonant frequency
of absorber

$$f_{res} = 55\sqrt{\left(\frac{a}{dV}\right)}$$

Figure 10.7 *Cavity absorber*

Practical absorbers

The acoustic tiles used on walls and ceilings often absorb sounds by a combination of several different methods, depending upon the frequency content of the sound. The basic material of the tile, such as fibreboard, is porous and acts as an absorbent for higher frequencies. The tile material may also be drilled with holes which then act as cavity absorbers.

Some tiles have a perforated covering, the holes in which form effective resonators. The tiles may also act as a panel absorber if they are mounted with an airspace behind them, such as in suspended ceiling. In general: panel absorbers are used for low frequencies; perforated panels are used for frequencies in the range 200 to 2000 Hz; and porous absorbers are used for high frequencies.

REVERBERATION

If the main source of sound in a room suddenly stops, it is unlikely that the sound in the room will also stop suddenly. A single hand clap demonstrates this effect. There is a continuing presence of sound, known as reverberation, which is particularly noticeable in a large reflective interior such as a church.

- ***REVERBERATION* is a continuation and enhancement of a sound caused by rapid multiple reflections between the surfaces of a room**

Reverberation is *not* the same as an echo as the reflections reach the listener too rapidly for them to be heard as separate sounds. Instead, the reverberative reflections are heard as an extension of the original sound.

When a pulse of sound is generated in an enclosed space the listener first receives sound in a direct path from the source. This direct sound is quickly supplemented by the sound reflected from the surfaces of the room. Some simplified reverberant sound paths are shown in figure 10.8. The sound loses some energy at each reflection, depending on the nature of the surface, and absorption usually limits the number of reflections causing reverberation.

If the original source of sound is continuous then the reverberant sound combines with the direct sound to produce a continuing *reverberant field*. A room totally without reverberation is termed *anechoic,* and is achieved by using special absorption techniques at each surface.

Reverberation time

If the source of sound stops then the reverberant sound level dies away with time, as shown in figure 10.9. The rate at which the sound decays is a useful

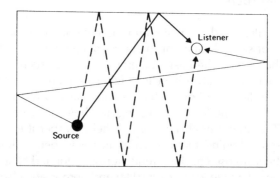

Figure 10.8 *Multiple reflections of reverberation*

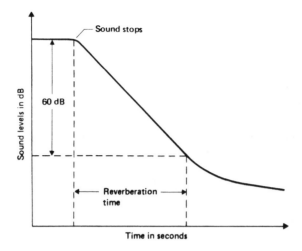

Figure 10.9 *Reverberation time*

indication of the reverberation quality and is measured by a reverberation time with a standard definition.

- ***REVERBERATION TIME*** **is the time taken for a sound to decay by 60 dB from its original level**

A decrease in sound level of 60 dB is the same as a drop to one millionth of the original sound power and represents the decay of a moderately loud sound to inaudibility. The time taken for this decay in a room depends upon the following factors:

- Areas of exposed surfaces
- Sound absorption at the surfaces
- Distances between the surfaces
- Frequency of the sound.

Reverberation time is an important index for describing the acoustical quality of an enclosure. The reverberation time of an existing auditorium can be determined by generating pulse sounds, such as gunshot from a starter's pistol, and measuring the decay time. The reverberation time of a planned auditorium can be calculated in advance from a knowledge of the factors above that affect reverberation time.

Ideal reverberation time

Typical reverberation times vary from a fraction of a second in small rooms to 5 seconds or more in very large enclosures like a cathedral. Different activities require reverberation times in the following ranges:

- **Speech:** 0.5 to 1 second reverberation time
- **Music:** 1 to 2 seconds reverberation time.

Short reverberation times are necessary for clarity of speech, otherwise the continuing presence of reverberant sound will mask the next syllable and cause the speech to be blurred.

Longer reverberation times are considered to enhance the quality of music which will otherwise sound 'dry' or 'dead' if the reverberation time is too short. Larger rooms are judged to require longer reverberation times, as is also the case with lower frequencies of sound.

Optimum reverberation times can be calculated by formulas, such as the following *Stephens and Bate* formula:

$$t = r\left[0.118^3\sqrt{(V)} + 0.107\right]$$

where t = reverberation time (s)
   V = volume of hall (m^3)
   r = 4 for speech, 5 for orchestras, 6 for choirs.

Ideal reverberation times can also be presented in sets of graphs, such as those shown in figure 10.10.

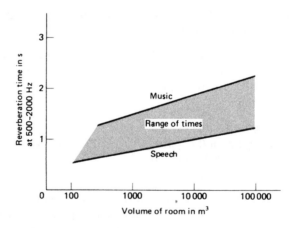

Figure 10.10 *Optimum reverberation times*

Reverberation time formulas

When reverberation time cannot be directly measured, as at the planning stage for example, it can be predicted from a knowledge of the factors that affect the decay of sound. Reverberation time depends on the areas of exposed surfaces, the absorption coefficients of these surfaces, and the distances between the surfaces (or volume). If these factors are numerically related by a formula then one of the factors can be calculated if the others are known.

Sabine's formula

Sabine's formula assumes that the reverberant decay is continuous and it is found to give reasonable predictions of reverberation time for rooms without excessive absorption.

$$t = \frac{0.16V}{A}$$

where t = reverberation time (s)
V = volume of the room (m^3)
A = total absorption of room surfaces (m^2 sabins)
= Σ (surface area $\times$ absorption coefficient).

Eyring formula

If the average absorption in a room is high, such as in a broadcasting studio, the reverberation times predicted by Sabine's formula do not agree with actual results. A more accurate prediction is given by the Eyring formula given below:

$$t = \frac{0.16V}{-S \log_e\left(1 - \bar{a}\right)}$$

where t = reverberation time (s)
V = volume of the room (m^3)
S = total area of surfaces (m^2)
$\bar{a}$ = average absorption coefficient of the surfaces.

Calculation of reverberation time

Reverberation times are calculated by finding the total absorption units in a room and then using a formula such as Sabine's formula. The absorption of materials varies with frequency and the reverberation time predicted by formula is only accurate for the frequency at which absorption coefficients are valid.

If the same value of reverberation time is required at different frequencies then the total absorption must be the same at each frequency. This can be achieved by choosing materials or devices that provide absorption at certain frequencies and not at others. For example, the absorption of porous materials increases with higher frequencies while the absorption of the panel decreases with higher frequencies.

Calculation guide

The worked examples illustrate common types of calculation and the following rules are useful:

- Do not directly add or subtract reverberation times with one another. Use Sabine's formula to convert reverberation times to absorption units, make adjustments by addition or subtraction of absorption units, then convert back to reverberation time.
- Surfaces that are not 'seen' do not usually provide absorption – an area of floor covered by carpet, for example.
- Use a sketch of the enclosure with dimensions to help identify all surfaces and their areas.
- Lay the calculations out in a table, as shown in the worked examples. This method encourages clear accurate working, allows easy checking, and is also suitable for transfer to a computer spreadsheet.

Worked example 10.1

A hall has a volume of $5000 \, m^3$ and a reverberation time of $1.6 \, s$. Calculate the amount of extra absorption required to obtain a reverberation time of $1 \, s$.

Know $t_1 = 1.6 \, s,$ $A_1 = ?$
$\quad\quad t_2 = 1.0 \, s,$ $A_2 = ?$
$\quad\quad V = 5000 \, m^3$

Using $t = 0.16 \, V/A$

for t_1

$$1.6 = \frac{0.16 \times 5000}{A_1} \quad \text{so} \quad A_1 = \frac{0.16 \times 5000}{1.6} = 500 \text{ sabins}$$

for t_2

$$1.0 = \frac{0.16 \times 5000}{A_2} \quad \text{so} \quad A_2 = \frac{0.16 \times 5000}{1.0} = 800 \text{ sabins}$$

Extra absorption needed $= A_2 - A_1 = 800 - 500$
$$= \textbf{300} \, m^2 \textbf{ sabins}$$

Worked example 10.2

A lecture hall with a volume of $1500 \, m^3$ has the following surface finishes, areas and absorption coefficients at $500 \, Hz$:

	Area	Abs coeff
Walls, plaster on brick	$400 \, m^2$	0.02
Floors, plastics tiles	$300 \, m^2$	0.05
Ceiling, plasterboard on battens	$300 \, m^2$	0.10

Calculate the reverberation time (for a frequency of 500 Hz) of this hall when it is occupied by 100 people.

Tabulate information and calculate the absorption units using

Absorption = Area × Absorption coefficient

Surface	Area m^2	500 Hz	
		Absorption coefficient	Abs. units (m^2 sabins)
Walls	400	0.02	8
Ceiling	300	0.10	30
Floor	300	0.05	15
Occupants	100 people	0.46 each	46
		Total A	99 sabins

Using Sabine's formula

$$t = \frac{0.16V}{A} = \frac{0.16 \times 1500}{99} = 2.42$$

So reverberation time = **2.42 s** at 500 Hz

Worked example 10.3

The reverberation time required for the hall in worked example 10.2 is 1.5 s. Calculate the area of acoustic tiling needed on the walls to achieve this reverberation time (absorption coefficient of tiles = 0.4 at 500 Hz).

The areas of tiles will change the original area of plain walls. The areas can be found by trial and error, or by algebra as shown here.

Surface	Area m^2	500 Hz	
		Absorption coefficient	Abs. units (m^2 sabins)
Tiles	S	0.40	0.4S
Walls	400−S	0.02	8−0.02S
Ceiling	300	0.10	30
Floor	300	0.05	15
Occupants	100 people	0.46 each	46
		Total A	99 + 0.38S

Using $t = \dfrac{0.16V}{A}$

$1.5 = \dfrac{0.16 \times 1500}{99 + 0.38S}$

Rearranging the formula

$0.38S = \dfrac{0.16 \times 1500}{1.5} - 99 = 160 - 99 = 61$

$S = \dfrac{61}{0.38} = 160.5$

So area of tiles = **160.5 m²**

Exercises

1 Draw a scaled plan and a section of a hall similar to that in figure 10.4, or use any suitable drawings of an auditorium. Choose a sound source situated on the centre of the stage and draw geometrically accurate sound path diagrams to show the reflections off the ceiling and off the walls. Comment on the distribution of sound in the hall and suggest remedies for any areas where reflections might cause acoustic defects.

2 A room of 900 m³ volume has a reverberation time of 1.2 s. Calculate the amount of extra absorption required to reduce the reverberation time to 0.8 s.

3 Calculate the actual reverberation time for a hall with a volume of 5000 m³, given the following data for a frequency of 500 Hz.

Surface area	Absorption coefficient
500 m² brickwork	0.03
600 m² plaster on solid	0.02
100 m² acoustic board	0.70
300 m² carpet	0.30
70 m² curtain	0.40
400 seats	0.30 units each

4 If the optimum reverberation time for the above hall is 1.5 s then calculate the number of extra absorption units needed.

5 A large cathedral has a volume of 120 000 m³. When the space is empty the reverberation time is 9 s. With a certain number of people present the

reverberation time is reduced to 6s. Calculate the number of people present, if each person provides an absorption of 0.46 m² sabins.

6 A rectangular hall has floor dimensions 30 m by 10 m and a height of 5 m. The total area of windows is 50 m². The walls are plaster on brick, the ceiling is hardboard on battens and the floor is wood blocks on concrete. There are 200 fabric seats. The reverberation time required for the hall, without audience, is 1.5 s at 500 Hz. Use table 10.1 to help calculate the area of carpet needed to achieve the correct reverberation time.

Answers

2 60 sabins

3 2.4 seconds

4 198.3 sabins

5 2319 people

6 71.1 m² carpet

11 Electricity Supplies

A supply of electricity is essential for creating and controlling the environment in a modern building. Systems for heating, cooling, ventilating, and lighting all use electricity, because of its energy content and for the ease with which electricity can be controlled. The electronic equipment of modern offices and intelligent buildings requires convenient supplies of electricity. On a larger scale, electricity also provides the large amounts of energy needed for pumping public water supplies.

This chapter explains the basic principles of electric currents and electric devices, such as generators and motors. These principles are then used to describe modern systems of generating and distributing electricity. The features of household electrical installations are also summarised.

CURRENT ELECTRICITY

Structure of matter

All material is made of atoms. An atom is the smallest part of matter that has a separate chemical existence. Atoms contain many smaller particles and among the forces that bind the sub-atomic particles together is an electric property called charge. There are two kinds of charges: positive (+) and negative (−). The forces between charges obey the following rules:

- **Like charges repel one another**
- **Unlike charges attract one another.**

There are three fundamental sub-atomic particles which help determine the nature of matter and give rise to electrical effects. The important properties of the fundamental sub-atomic particles are shown in figure 11.1 and summarised below.

Protons
- Protons have a positive electric charge, equal and opposite to that of the electron

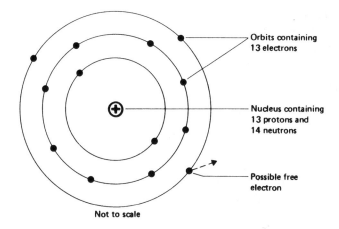

Not to scale

Figure 11.1 *Aluminium atom, simple model*

- Protons have a mass of 1 atomic mass unit
- Protons are found in the central nucleus.

Neutrons
- Neutrons have no electric charge so are 'neutral'
- Neutrons have the same mass as the proton
- Neutrons are found in the nucleus with the proton.

Electrons
- Electrons have a negative electric charge, equal and opposite to that of the proton
- Electrons have very small mass, approximately 1/1840 atomic mass unit
- Electrons are found surrounding the nucleus.

The nucleus occupies a very small volume at the centre of the atom but contains all the protons and neutrons. Therefore, despite its small size, the nucleus contains nearly all the mass of an atom. The electrons can be considered as circulating in orbits around the nucleus, held in position by the opposing charge of the protons in the nucleus. An atom contains the same number of electrons as protons, so the positive and negative charges are balanced and the overall charge of an atom is zero.

The electrons in the outer orbits are held by relatively weak forces so outer orbits can sometimes lose or gain electrons. *Free electrons* are electrons in outer orbits that are able to move from one atom to another. *Ionisation* is a process in which an atom permanently gains or loses electrons and so acquires an overall charge. This charged atom is then called an *ion*.

Electric charge

When electrons move from one place to another, by whatever mechanism, they transfer their electric charge. It is this charge that electricity is composed of.

- **CHARGE (Q) is the basic quantity of electricity**

  UNIT: coulomb (C)

The idea of charge or static electricity is more fundamental than current electricity. Although current electricity is more usual in everyday applications, electricity does not have to move to exist. A thundercloud, for example, contains a huge quantity of electricity, which does not flow, except during a stroke of lightning.

Electric current

If electric charge transfers through a material then an electric current is said to 'flow'. The movement of free electrons is the usual mechanism for the transfer of charge. The amount of electric current is described by the quantity of charge which passes a fixed point in a given time.

- **ELECTRIC CURRENT (I) is the rate of flow of charge in a material**

  UNIT: ampere (A)

This relationship is written as the following formula:

$$I = \frac{Q}{t}$$

where I = electric current flowing (A)
   Q = electric charge (C)
   t = time taken (s).

Direction of current

Electrons have a negative charge and, by the rules of charge, they are attracted to a positive charge. Therefore when electrons move through a material, such as a cable, they are attracted to the positive side of the electrical supply, as shown in figure 11.2. However, by convention, it is usual to say that direct electric current flows from positive to negative, even though electrons actually flow in the opposite direction. This convention works for practical problem solving as long as it is maintained consistently.

Effects of current

Electrons are too small to be detected themselves, but the presence of an electric current can be known by its effects. Three important effects are listed below:

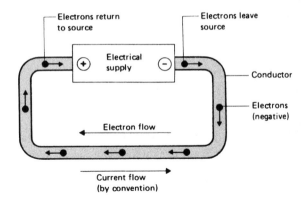

Figure 11.2 *Electron and current flow*

- **Heating effect:** Current flowing in a conductor generates heat. The amount of heat depends upon the amount of current but does not depend upon its direction. Applications of the heating effect include electric heaters and fused switches.
- **Magnetic effect:** Current flowing in a conductor produces a magnetic field. The size and direction of the field is affected by the size and direction of the current. An electric motor is one example of this effect.
- **Chemical effect:** Current flowing through some substances causes a chemical change and produces new substances. The type of change depends upon the amount and direction of the current. This type of effect is used in electroplating and refining processes.

Conductors and insulators

If a material allows a significant flow of electric current then the material is termed a conductor of electricity. A material that passes relatively little current is termed an insulator.

Conductors
Solid conductors are materials whose free electrons readily produce a flow of charge. If the conductor is a liquid or a gas then the charge is usually transferred by the movements of ions. Common types of conductor are listed below:

- **Metals:** Examples include copper and aluminium cable conductors
- **Carbon:** Examples include sliding contacts in electric motors
- **Liquids and gases:** Current can flow when ions are present, such as in salty water or in a gas discharge lamp.

Insulators

Insulators are materials that have few free electrons available to produce a flow of charge. Common types of insulator are listed below:

- **Rubber and plastic polymers:** Examples include PVC cable insulation
- **Mineral powder:** Examples include magnesium oxide cable insulation (MICC)
- **Oil:** Examples include underground cable insulation
- **Dry air:** Examples include overhead power line insulation
- **Porcelain and glass:** Examples include overhead power line insulation.

Electrical potential

Potential difference

For an electric current to flow in a conductor there must be a difference in charge between two points. This potential difference is similar to the pressure difference that must exist for water to flow in a pipe.

- ***POTENTIAL DIFFERENCE (pd or V) is a measure of the difference in charge between two points in a conductor***

  UNIT: volt (V)

The volt is defined in terms of the energy needed to move an electric charge. Because potential difference is measured in volts it is sometimes termed 'voltage drop'. Current flows from the point of higher potential to the point of lower potential, as shown in figure 11.3.

Electromotive force

In order to produce a potential difference and the resulting current, there must be a source of electrical 'pressure' acting on the charge. This source, shown in figure 11.3, is called an electromotive force (EMF).

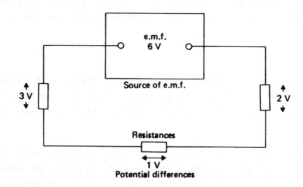

Figure 11.3 *Potential difference and electromotive force*

- An *ELECTROMOTIVE FORCE (E)* is a supply of energy capable of causing an electric current to flow

UNIT: volt (V)

Because the unit of EMF is the volt, the same unit as for potential difference, the EMF of a circuit is sometimes called the 'voltage'. Common sources of EMF are batteries and generators.

Resistance

Some materials oppose the flow of electric current more than others. Resistance is a measure of opposition to the flow of electric current and is related to the potential difference and the current associated with that opposition.

$$\textbf{RESISTANCE } (R) = \frac{\text{Potential difference } (V)}{\text{Current } (I)}$$

UNIT: ohm (Ω) where $1\,\Omega = 1\,\text{V/A}$

A *resistor* is a component that is used to provide resistance and its resistance depends upon the following factors:

- **Length:** Doubling the length doubles the resistance.
- **Cross-sectional area:** Doubling the area decreases the resistance by half.
- **Temperature:** The resistance of most metals increases with temperature.
- **Material:** The resistance provided by a particular material is given by a value of *electrical resistivity*. The resistivity of copper, for example, is about 1/2 that of aluminium, about 1/6 that of iron, and about 10^{-20} that of a typical plastic.

Ohm's law
The potential difference and the current associated with a resistance may be measured and the values compared. For most metal conductors it is found that if, for example, the potential difference (voltage) is doubled then the current also doubles. This relationship is expressed as Ohm's Law:

- *OHM'S LAW:* **For a metal conductor, at constant temperature, the current flowing is directly proportional to the potential difference across the conductor**

The constant of proportionality is, by definition, the resistance of the conductor; so that the following useful expression is a result from Ohm's Law:

$$V = IR$$

where V = potential difference across a component (V)
I = current flowing in the component (A)
R = resistance of the component (Ω).

Note that this expression can also be expressed in two other forms as follows:

$$I = \frac{V}{R} \quad \text{and} \quad R = \frac{V}{I}$$

Worked example 11.1

An electronic calculator has an overall resistance of 6 kΩ and is connected to a supply which provides a potential difference of 9 V. Calculate the current flowing to the calculator.

Know $V = 9\text{V}$ $I = ?$ $R = 6000\,\Omega$

Using

$$V = IR$$

$$9 = I \times 6000$$

$$I = \frac{9}{6000} = 1.5 \times 10^{-3}\text{A}$$

So current = **1.5 mA**

Circuits

For a continuous electric current to be able to flow there must be a complete circuit path from, and back to, the source of electromotive force. In most circuits this complete path is supplied by two obvious conductors connecting the electrical device to the supply. Some systems, however, use less obvious means, such as the conduction of a metal structure or of the earth itself as one part of the circuit. For example, the metal chassis of a car or a television set is part of the circuit; and the earth itself forms part of the circuit for the distribution of electrical energy to buildings.

For simple problems the resistance of the connecting conductors is assumed to be negligible. The voltage drop produced by the resistance of practical lengths of cable can be predicted from published tables and, if it is excessive, a cable of lower resistance is used. The theoretical layout of circuits is shown in a geometrical manner using standard symbols, as used for the diagrams in this chapter. In an actual wiring system the connections are made on the components, rather than on the conductors as shown in the circuit diagrams.

Circuit rules

There are basic principles regarding the behaviour of current, voltage, and resistance that apply to any circuit or part of a circuit. If some of the electrical values are known for a circuit then others can be predicted by the use of these rules.

There are two basic layouts for connecting components in a circuit: parallel and series connection. Figures 11.4 and 11.5 show the two types of circuit using three components, but the circuit rules apply to any number of components.

Series connection
- **Current:** The current flowing in each resistor has the same value
- **Voltage:** The sum of the voltage drops across all the resistors is equal to the applied voltage

$$V = V_1 + V_2 + V_3$$

- **Resistance:** The total resistance of the circuit is equal to the sum of the individual resistances

$$R = R_1 + R_2 + R_3$$

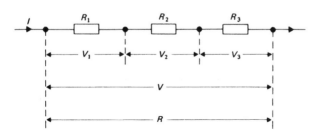

Figure 11.4 *Resistors connected in series*

Parallel connection

- **Current:** The sum of the currents in all the resistors is equal to the total current flowing in the circuit

$$I = I_1 + I_2 + I_3$$

- **Voltage:** The voltage drop (potential difference) across each resistor has the same value
- **Resistance:** The total resistance of the circuit is obtained from the reciprocals of the individual resistances

$$\frac{1}{R} = \frac{1}{R_1} + \frac{1}{R_2} + \frac{1}{R_3}$$

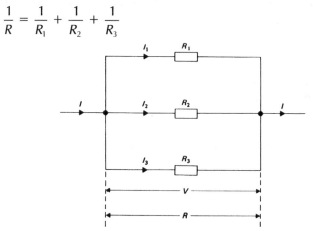

Figure 11.5 *Resistors connected in parallel*

Worked example 11.2

A 3 Ω and a 6 Ω resistor are connected together in parallel and then connected in series with a 4 Ω resistor. If an EMF of 24 V is applied to this circuit then calculate the following items:

(a) The total resistance of the circuit.
(b) The total current flowing in the circuit.
(c) The potential difference across the 3 Ω resistor.
(d) The current flowing in the 3 Ω resistor.

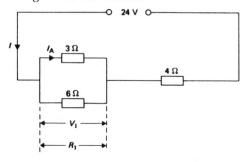

Figure 11.6 *Worked example 11.2*

Draw the circuit and label the known and unknown values, as in figure 11.6.

(a) Let total resistance be R_T

$$\frac{1}{R_1} = \frac{1}{3} + \frac{1}{6} = \frac{2+1}{6} = \frac{3}{6} \quad \text{so} \quad R_1 = 2\Omega$$

$R_2 = 4\Omega$

So $R_T = R_1 + R_2$
   $= 2 + 4 = 6\Omega$

(b) Let total current be I_T
   Total $R = 6\Omega$, Total $V = 24\,V$

$$I = \frac{V}{R} = \frac{24}{6} = 4A$$

(c) Potential difference V_1
   Resistance $R_1 = 2\Omega$, Current in $R_1 = 4\,A$

$V_1 = I \times R_1$
   $= 4 \times 2 = 8\,V$

(d) Current I_A

$$I_A = \frac{V_1}{R} = \frac{8}{3}$$
   $= 2.67\,A$

Cells

A cell is a device that converts chemical energy to electrical energy and supplies an electromotive force (EMF) capable of causing a direct electric current to flow. There are two classes of cell: primary and secondary cells.

Primary cells
   • **Primary cells** are cells that cannot be recharged.

In a primary cell, the conversion of energy is not reversible and the cell must be replaced. Examples of these cells include simple cells, zinc–carbon cells, dry mercury cells, alkaline manganese cells and silver oxide cells.

Secondary cells
   • **Secondary cells** are cells that can be recharged.

In a secondary cell, the conversion process can be reversed, the energy content replaced, and the cell can be used again. Examples of secondary cells include lead–acid accumulators (car batteries) and nickel–cadmium cells.

Types of cell

Some common forms of cell are described below and all of them have the same basic components of *electrodes* and *electrolyte*.

- **Electrodes:** Electrodes are conductors which form the terminals of the cell. The *anode* is the positive electrode and the cathode is the negative electrode.
- **Electrolyte:** The electrolyte is a compound that undergoes chemical change and releases energy.

Simple cell

A simple or 'voltaic' cell is a primary cell which is constructed as shown in figure 11.7. This cell is not a practical source of supply as it has a limited life, mainly because of polarisation. *Polarisation* is a reverse EMF set up in the cell because hydrogen is liberated and deposited on the copper electrode. The general nature of a simple cell is an important mechanism in *electrolytic corrosion* when simple cells can occur between neighbouring areas of different metals.

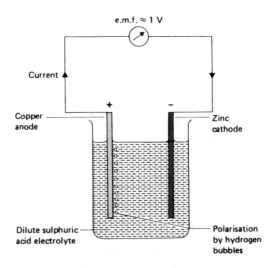

Figure 11.7 *Simple cell*

Dry cell

The dry cell or zinc–carbon cell is the commonest type of primary cell and its construction is shown in figure 11.8. The dry cell is a form of

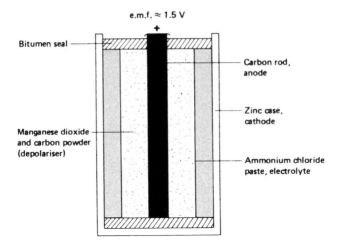

Figure 11.8 *Dry cell*

Leclanché cell with the electrolyte made as a paste rather than a liquid. The nominal EMF of any single dry cell is 1.5 V and the cell contains a depolarising agent. It is suitable for intermittent use in torches, radios, and similar devices.

Lead–acid cells

The lead–acid cell is the commonest type of secondary cell and is used in most motor vehicles. The electrodes are made of lead and lead oxide set in lead alloy grids; the electrolyte is dilute sulphuric acid. Each electrode changes from one form of lead to the other as the cell is charged or discharged; the concentration of the acid electrolyte indicates the state of charge. The EMF of any single lead–acid cell is 2 V.

Connections between cells

A *battery* is a combination of more than one cell connected together to give an increased EMF or increased capacity.

- **Series connection:** This type of connection is shown in figure 11.9. The total EMF of the battery is the sum of the individual EMFs.

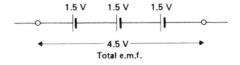

Figure 11.9 *Cells connected in series*

- **Parallel connection:** This type of connection is shown in figure 11.10. The total EMF of the battery is equal to the EMF of a single cell. The maximum current or the life of the battery is three times that of a single cell.

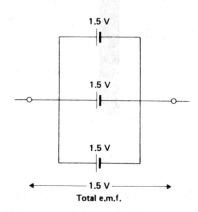

Figure 11.10 *Cells connected in parallel*

Power and energy

Electrical energy and power have the same meaning and the same units as other forms of power and energy. Power is the rate of using energy and, because the volt is defined in terms of electric charge and energy, it is possible to express power in terms of electric current (flow of charge) and voltage.

Electrical power
POWER (*P*) = Current (*I*) × Potential difference (*V*)

$$P = IV$$

UNIT: watt (W)
where, by definition, 1 watt = 1 joule/second

Two other useful expressions are obtained by combining the expressions $P = IV$ and $V = IR$:

$$P = I^2R \quad \text{and} \quad P = \frac{V^2}{R}$$

The *power rating* of an electrical appliance is often quoted in specifications and it is an indication of the relative energy consumption of the device. Some typical power ratings are given below:

electric kettle element   2500 W
electric fire (1 bar)   1000 W
colour television   100 W
reading lamp   60 W
calculator charger   5 W

Electrical energy
Power is defined by the energy used in a certain time, so it is also possible to express energy in terms of power and time:

ENERGY (E) = Power (P) × Time (t)

$$E = Pt$$

UNIT: joule (J)
where 1 joule = 1 watt × 1 second

The *kilowatt hour* (kWh) is an alternative unit of energy in common use for electrical purposes, where

1 kWh = 1 kilowatt × 1 hour = 3.6 MJ

Worked example 11.3
A 3 kWh electric heater is connected to a 240 V supply and is run continuously for 8 hours. Calculate: (a) the current flowing in the heater; and (b) the total energy used by the heater.

(a) Know $P = 3000$ W $I = ?$ $V = 240$ V

Using $P = IV$

$$3000 = I \times 240$$

$$I = \frac{3000}{240} = 12.5$$

So current = **12.5 A**

(b) Know $E = ?$ $P = 3$ kW $= 3000$ W $t = 8$ h $= 28\,800$ s.

Using $E = Pt$

$$E = 3000 \times 28\,800$$
$$= 86.4 \times 10^6 \text{ J}$$

Alternatively, using kilowatt hours

$E = 3 \times 8$
$ = 24\,kWh$

So energy $= $ **86.4 MJ** or **24 kWh**

MAGNETISM

There are certain pieces of iron, and other materials, that can push or pull on one another when there is no physical contact between them. This force of magnetic attraction or repulsion is caused by the movement of charged particles inside the material. This motion may be due to the natural spin of sub-atomic particles, like the electron, or it may be due to the flow of electrons in an electric current. Magnetism is therefore linked with electric currents. This section describes those effects where electricity and magnetism combine to produce forces which can be used in devices such as motors.

Magnetic fields

The individual atoms of all materials have magnetic properties but they are not usually detected outside the atoms. In certain materials the magnetic effects of the individual atoms can be aligned to produce an overall effect. Iron, nickel and cobalt, for example, can retain this magnetic alignment and are used to make permanent magnets.

A *magnetic field* is a region where a magnetic force can be detected. The common sources of magnetic fields include the following:

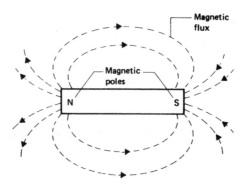

Figure 11.11 *Magnetic field of bar magnet*

- Permanent magnetic materials, such as iron
- A conductor carrying an electric current
- The interior of some planets, such as Earth.

A simple *compass* is made from a magnet balanced on a string or a pivot. In the presence of a magnetic field the compass will turn and align itself in the direction of the magnetic field. The direction of the magnetic field changes with position and it is useful to consider this pattern of the magnetic field as being made up of 'lines of force' or *magnetic flux.*

The magnetic field pattern for a simple bar magnet, in one plane only, is shown in figure 11.11. The lines of magnetic flux begin and terminate at the two *poles* of the magnet: a 'north-seeking pole' and a 'south-seeking pole'. By convention, the magnetic field is said to flow from north pole to south pole.

When two magnets are placed close together the forces between them act towards or away from the poles of the magnets. The direction of the force follows the general rules:

- Unlike poles attract one another
- Like poles repel one another.

Two north poles, for example, will push away from one another; while a north and a south pole will pull towards one another.

Electromagnetism

A conductor carrying an electric current produces a magnetic field around itself. A compass placed near a wire shows that the field is in the form of concentric circles; in a clockwise direction for current travelling away from view. A single turn of a loop of wire sets up two magnetic fields which combine, in the same direction, between the loops.

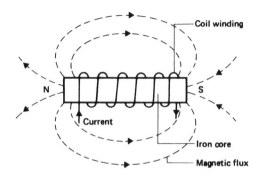

Figure 11.12 *Magnetic field of solenoid*

A *solenoid* is coil made up of many loops of wire. When a current flows in a solenoid the magnetic field produced around the solenoid, shown in figure 11.12, is found to be similar to that of a bar magnet, with north and south poles produced at the ends of the solenoid.

An *electromagnet* is the general name given to a solenoid that is connected to an electric supply. If the solenoid is wound on an *armature* core of iron, the lines of magnetic flux are concentrated and change the shape of the magnetic field.

Applications of electromagnets

The main property of an electromagnet is its dependence on the supply of current: it can be turned on and off. Electromagnets are used instead of permanent magnets as a source of magnetic fields. Some important applications of electromagnets are given below:

- Lifting iron and steel, such as in scrap metal yards
- Solenoid valves for turning supplies of gas and water on
- Electric bells and buzzers

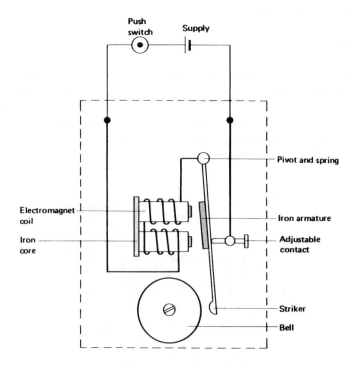

Figure 11.13 *Electric bell*

- Relay switches for controlling circuits at a distance
- Circuit breakers for preventing excessive current flows.

Electric bell

An electric bell or buzzer uses an electromagnet as shown in figure 11.13. Pressing the bell-push connects the bell to the supply and magnetises the electromagnet which then attracts the armature. But the movement of the armature breaks the circuit at the contact points and the magnetism dies away so that the armature springs back to its original position, completes the circuit, and starts the cycle again. The continuous trembling of the armature can be used to ring a bell, or to produce a buzzing sound, for as long as the switch is pushed.

Relay

A relay switch is a device used for controlling one circuit by means of another circuit. A relay uses the force produced by an electromagnet and a simple example is shown in figure 11.14. Switching on the control circuit magnetises the electromagnet which then attracts the armature. The movement of the armature closes the contacts of the relay switch and causes the main circuit to be turned on.

The advantage of this type of switch is that the control circuit can use low voltage and current to control a main circuit which carries a larger load and is situated some distance away.

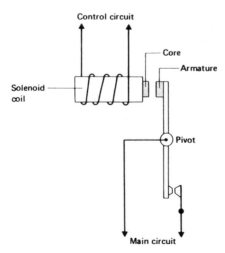

Figure 11.14 *Principle of relay switch*

Force on a conductor

When a conductor in a magnetic field carries a current, a force is found to act on that conductor. This *motor effect* is caused by the interaction of two magnetic fields: the field due to the magnet and the field due to the current.

The strength of the force on the conductor is increased by the following measures:

- Increase in current
- Increase in magnetic field strength
- Increase in length of the conductor sited in the magnetic field.

If the conductor in the magnetic field is in the form of a loop or a coil then a force acts on each side of the coil. These forces act in opposite directions and they combine to produce a turning effect, called a torque, on the coil. The turning action of a current-carrying coil in a magnetic field is the basis of devices such as direct current motors and electric meters.

Electric motors

An electric motor is a device for converting electrical energy to mechanical energy by using the force produced when two magnetic fields interact.

Simple DC motor
The simple motor illustrated in figure 11.15 is a design that works but would not be used for a practical machine. However, the essential features and the operation of a simple motor are convenient to describe and they also apply to practical motors. The coil is set in a magnetic field and is free to turn on an axle. The DC supply is connected to the coil via conducting 'brushes' which slide against the rotating *commutator*.

When current flows in the coil the motor effect produces forces that turn the coil. When the coil is vertical the commutator disconnects the supply and the coil continues turning by momentum. The commutator then reconnects the coil to the opposite supply terminals. This reversal of connections ensures that the same direction of force always acts on the same side of the motor. The repeating cycle of reversed connections keeps the axle turning in one direction and it can be used to do mechanical work.

Practical DC motors
A practical motor works on the same general principle as the simple motor but is made more efficient by additional features.

Field windings supply the magnetic field. These electromagnets are arranged in a circular form in order to keep the air gap to a minimum.

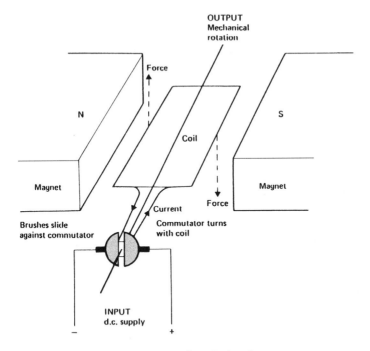

Figure 11.15 *Simple DC electric motor*

The *armature* turns on an axle inside the magnetic field and has the main coils wound upon it. More than one set of coils is wound on the same armature and a commutator, with many splits, connects the electrical supply to one coil at a time.

The connections between field coils, the armature windings, and the electrical supply can be interconnected in different ways, which causes the motor to show different characteristics when starting and when under load. The speed of the motor can be varied by the use of a variable resistor in series with the armature.

AC motors

Most household electric motors, and many industrial motors, operate from an alternating current supply. The magnetic interactions of such motors involve the properties of induction, discussed in the next section. In general, the effect of alternating current is to set up rotating magnetic fields in the stator (field windings). The magnetic field of the rotor (armature) tends to follow the rotating fields and so causes the motor to turn. The construction of AC motors can be simple but speed control of AC motors is more difficult than for DC motors.

INDUCTION

One general meaning of the word 'induction' is the production of an effect without physical contact. It is possible for an electrical effect to be induced in one circuit by the action of another circuit, even though there is no apparent contact between the two circuits. In fact there is always a very real magnetic link of magnetic flux. Induction is a special form of electromagnetic effect which explains the operation of important features of public electricity supplies such as generators, transformers, and alternating current properties.

Electromagnetic induction

If a conductor is moved across a magnetic field then an EMF is induced in that conductor. This induced electricity also occurs in the types of circuit shown in figure 11.16.

Moving magnet
In figure 11.16a, a magnet is moved towards the coil and an EMF is induced in the coil. A current therefore flows in the circuit and is detected by the meter. The direction of the current reverses when the direction of the magnet

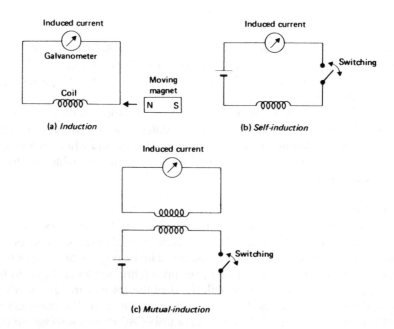

Figure 11.16 *Electromagnetic induction*

is reversed or the poles of the magnet are reversed. The induction occurs when either the coil or the magnet moves.

Switching circuit
If the coil and the meter are connected directly to a DC supply, as in figure 11.16b, there is no induced EMF. If, however, a switch is inserted in the circuit then the EMF changes every time that the switch is put either on or off. This *self-induction* is produced when the electromagnetic field of the coil cuts across adjoining turns of the coil.

Adjacent coils
If two coils are placed adjacent to one another, as in figure 11.16c, then an EMF is induced in the meter whenever the switch in the battery circuit is put either on or off. This *mutual inductance* is produced when the electromagnetic field of the first coil cuts the turns of the second coil. The 'flux linkage' between the two coils can be improved by using a soft-iron core between them.

Principles of induction

The different demonstrations of induction described above share a common mechanism – in each case, the induction coil experiences a magnetic field which is changing. The change in magnetic field may be caused by a movement or by switching a circuit on or off.

General principle of induction
- **An electric current will be induced in a conductor which is subjected to a *changing* magnetic field**

Magnitude of induction
The size of the induced current depends upon the following factors:

- Relative speed of movement of the magnetic field
- Strength of the magnetic field
- Length of conductor in the magnetic field
- Angle between the conductor and the field.

Lenz's law
The direction of all induced currents can be predicted by Lenz's Law, given below:

- **LENZ'S LAW: the direction of the induced current is such that it will always oppose the change that produced it**

This rule applies to all methods of induction. In the simple example shown in figure 11.16a a magnet is moved towards the coil with the north pole leading. The current induced in the coil sets up a magnetic field and, by Lenz's Law, this field will oppose the approaching magnet by having its own north pole outwards. If the movement of the magnet is reversed then the induced field will also be reversed; so as to oppose the movement.

Applications of induction

Electromagnetic induction is the principle behind many important devices used for the generation, the transmission, and the application of electricity. Some examples are given below:

- Generators: for the production of electricity
- Transformers: for changing voltage
- Induction motors
- Ignition coils: for spark in car engines
- Hi-Fi cartridges: for playing records
- Linkages in electronic circuits.

Generators

A generator is a device that converts mechanical energy to electrical energy by means of electromagnetic induction. Most of the electricity used in everyday life is generated in this manner. The change in magnetic field necessary for induction is produced by moving a coil through a magnetic field, or by moving a magnetic field past a coil. Rotational motion is usually employed. Some types of generator may be known as a dynamo or an alternator.

Simple AC dynamo

The simplest type of generator is the AC dynamo, shown in figure 11.17. The generator coil is set on an axle in a magnetic field and connected to metal slip-rings on the axle. The slip-rings make sliding contact with the stationary brushes, which carry current away from the generator. The axle is rotated by a source of mechanical energy, such as a motor, and the coil moves through the magnetic field. The coil then experiences a changing magnetic field so that an EMF is induced in the coil, and therefore a current flows. The current is led away from the generator via the slip-rings and the brushes.

Maximum current is generated twice per revolution as the coil cuts the field at right angles, which happens when the coil is horizontal. Zero current is generated when the coil is vertical. The current varies from zero to maximum depending upon the angle at which the coil cuts the magnetic field. As the coil rotates through one revolution it cuts the magnetic field in two different directions and the induction reverses direction.

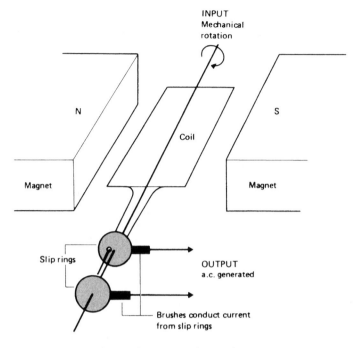

Figure 11.17 *Simple AC dynamo*

The EMF and the current produced by this generator are therefore continuously 'alternating' with every revolution of the coil. The output of this 'alternator' is in the shape of a sine wave, as shown in figure 11.19.

DC generator

The current from a battery has a steady value in one direction. To obtain this sort of 'direct current' from a generator it is necessary to replace the slip-rings by a commutator which reverses the connections every half turn and gives a current flowing in one direction only. This current will still fluctuate in value and to obtain a virtually steady flow of current the commutator is split into many sections and connected with up to 30 different coils. The construction is the same as for the DC motor and some machines can act as both generator and motor, at different times.

Practical AC generators

A large practical AC generator or alternator works on the same general principle as the simple generator but its construction is different.

The *stator* is the stationary frame to which are fixed the output windings that produce the current. No moving contacts are necessary to lead the current from the generator so the construction is suitable for large supplies.

The *rotor* turns inside the stator and contains the magnetic field coils. The DC supply necessary for these electromagnets is supplied by a separate self-starting *excitor* dynamo, run on the same axle.

Transformers

A transformer is a device that uses the principle of electromagnetic induction for the following purposes:

- To step up or step down voltage
- To isolate a circuit from an AC supply.

A major reason for using AC in electricity supplies is the relative simplicity and efficiency of transformers for changing voltage, as explained in the section on power transmission. A transformer does not work on DC supply. The construction of a common type of transformer is shown in figure 11.18.

Operation of transformer
The primary coil is connected to an AC supply and sets up a magnetic field which is continuously changing with the alternating current. The secondary coil experiences this changing magnetic field and produces an induced EMF which can then be connected to a load. The induced EMF alternates with the same shape and frequency as the EMF of the supply frequency but the ratio of the two EMFs is proportional to the ratio of the turns on the two windings.

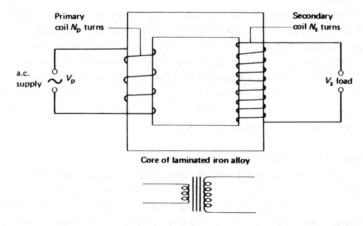

Figure 11.18 *Double-wound step-up transformer*

Transformer equation

$$\frac{V_s}{V_p} = \frac{N_s}{N_p}$$

where V_s = EMF induced in the secondary coil (V)
V_p = EMF applied to the primary coil (V)
N_s = number of turns on the secondary coil
N_p = number of turns on the primary coil.

A transformer with twice the number of secondary turns as primary turns will, for example, 'step up' the secondary voltage to twice the primary voltage. If the connections to such a transformer are reversed the transformer could be used as a 'step down' transformer which then halves the applied voltage. In an 'isolating transformer' the primary and secondary voltage may be the same.

Transformers are efficient machines and their power output is almost equal to their power input. If, for simplicity, it is assumed that there are no power losses then the following relationship is true:

Output power = Input power

For the practical case where the transformer is less than 100 per cent efficient, the following formula is used:

$I_s V_s = I_p V_p \times$ efficiency factor

where I_s = current in secondary coil (A)
V_s = EMF of secondary coil (V)
I_p = current in primary coil (A)
V_p = EMF of primary coil (V).

This relationship shows that if a transformer increases the voltage then the current must decrease by the same ratio, so as to conserve energy. In practice, some energy is lost in a transformer by heating in the coils and in the core. *Eddy currents* are circulating currents which are induced in the core. These currents are minimised by constructing the core from separate laminations so that core presents a high magnetic and electrical resistance.

Worked example 11.4
A transformer has 600 turns on the primary coil and 30 turns on the secondary coil. An EMF of 240 V is applied to the primary coil and a current of 250 mA flows in the primary coil when the transformer is in use.

(a) Calculate the EMF of the secondary coil.
(b) Calculate the current flowing in the secondary coil. Assume that the transformer is 95 per cent efficient.

(a) Know $N_p = 600$ $V_p = 240$ $N_s = 30$ $V = ?$

Using

$$\frac{V_s}{V_p} = \frac{N_s}{N_p}$$

$$\frac{V_s}{240} = \frac{30}{600}$$

$$V_s = \frac{30}{600} \times 240 = 12\,V$$

So secondary EMF = **12 V**

(b) Know $V_p = 240\,V$ $I_p = 250/1000 = 0.25\,A$ $V_s = 12\,V$ $I = ?$

Using $I_s V_s = I_p V_p \times 95/100$

$$I_s \times 12 = 0.25 \times 240 \times 95/100$$

$$I_s = \frac{0.25 \times 240 \times 95}{12 \times 100} = 4.75\,A$$

So secondary current = **5 A**

Alternating current properties

The EMF induced in an AC generator is constantly changing and reversing and so the current produced by the EMF also changes to give a pattern shown in figure 11.19. Some additional terms are needed to describe the nature of the alternating output.

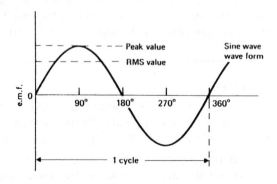

Figure 11.19 *Alternating current output*

Frequency

- *FREQUENCY (f)* **is the number of repetitions, or cycles, of output per second**

  Unit: hertz (Hz)

For public supplies in Britain and other countries in Europe the frequency is 50 Hz; in North America the frequency is 60 Hz.

Peak value

The peak value is the maximum value of alternating voltage or current, measured in either direction. The peak values occur momentarily and only twice in a complete cycle, as shown in figure 11.19.

RMS value

The simple mathematical average of a sine wave output is zero. But an alternating supply does produce an effective voltage or current and this is measured by RMS values.

- **A** *ROOT MEAN SQUARE (RMS)* **value of alternating current is that value of direct current that has the same heating effect as the alternating current**

A 1 kW fire, for example, produces the same heating effect using 240 V alternating current as it does using 240 V direct current. The RMS value is found mathematically by taking many instantaneous peak values, squaring them, taking the average of the squares, then taking the square root. The relationship between the values is found to be

RMS = 0.707 peak

The domestic supply AC voltage of 240 V RMS has a peak value of 339 V. The values of EMF and current quoted for AC supplies are assumed to be RMS values, unless it is stated otherwise.

Power factor

The voltage and the current of an AC supply are *in phase* when both have their peak values and zero values occurring at the same time. The power used by an AC circuit that is in phase is calculated as the product of the RMS values of current and voltage; similar to DC power.

Some AC circuits contain components that cause a *phase shift*, where either the voltage or the current leads or lags the other. In such a circuit some of the energy supplied as *apparent power* is lost in heating the circuit and does not appear as effective power or *active power*. The power factor expresses the ratio between the two forms of power:

$$\textbf{POWER FACTOR (PF)} = \frac{\text{Active power (watts)}}{\text{Apparent power (volt-amperes)}}$$

$$= \text{cosine } \theta$$

where θ is the phase angle between current and voltage.

The power factor has a maximum value of 1. Most AC equipment is rated by its apparent power together with a power factor. Supply authorities have to generate the apparent power but can only charge for the active power which appears on the meters, so they set minimum allowable values for the power factor (0.85 for example).

Devices with simple resistance, such as a heater, have a power factor of 1 and their active power is the same as their apparent power. Devices with induction coils, such as motors for example, have power factors less than 1. Low power factors can be improved by adding a capacitance device in parallel to balance the inductance.

Three-phase supply

The output from a simple AC generator, shown in figure 11.19, is a *single-phase* supply. Practical supplies of electricity are usually generated and distributed as a *three-phase* supply, which is – in effect – three separate single-phase supplies each equally out of step, as shown in figure 11.20. The generator is essentially composed of three induction coils instead of one, each coil being displaced from one another by 120 degrees. The three phases can be interconnected in different ways for different purposes, as shown in figure 11.21.

A *delta* connection is used for larger loads, such as three-phase motors. The voltage across any two phases (the line voltage) is 1.73 ($\sqrt{3}$) times the voltage between any one phase and earth.

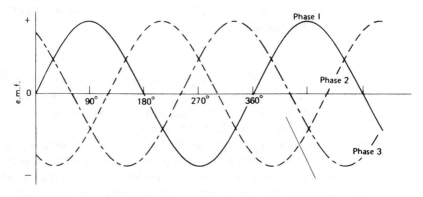

Figure 11.20 *Three-phase alternating current*

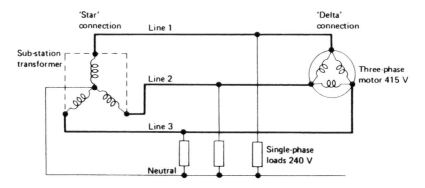

Figure 11.21 *Three-phase, four-wire AC supply*

A *star* connection is used for relatively small loads, such as households. They are connected to one phase and to the neutral to provide a 240 V supply. If the single-phase loads are evenly balanced then the return current in the neutral cable is zero.

Three-phase supplies are economical in their use of conductors and can supply more power than single phase supplies. Only three cables need to be used for the three-phase supply, instead of the six cables needed for three separate single-phase supplies. A device such as a motor which is designed to make use of the three phases receives more energy per second than those devices wired to a single phase; the three-phase motor is also smoother in its operation.

POWER SUPPLIES

A public supply of electrical energy is one of the most important services in the built environment. Manufacturing, building services, transport, and communications are all dependent upon electricity supplies. Large electric power systems require significant investments in construction. They also have a considerable impact upon the appearance of the landscape and any breakdown in the system greatly disrupts everyday life. Patterns of electrical energy use in buildings will tend to change in the future and the design of electric power systems will also need to evolve.

A complete power system is a collection of equipment and cables capable of producing electrical energy and transferring it to the places where it can be used. A power system is made up from the following three main operations:

- **Generation:** the production of the electricity
- **Transmission:** the transfer of electrical energy over sizeable distances

- **Distribution:** the connection of individual consumers and the sale of electricity.

The overall energy efficiency of a power system is about 30 per cent.

Power stations

All power stations generate electrical energy by using electromagnetic induction where an EMF is produced in a coil which experiences a changing magnetic field. The mechanical energy required is obtained using the heat from burning fossil fuels and from nuclear reactions, or obtained from the energy of moving water. A turbine is a device that produces rotational motion from the steam or running water and turns the axle of the generator.

A typical generator in a thermal power station turns at 3000 revolutions per minute and has an excitor dynamo mounted on the same shaft. The output varies but a large 500 MW generator set commonly generates 23 000 A at 22 kV. The cooling of the generators, by liquid or gas, is an important part of their engineering.

Thermal power stations

Thermal power stations use heat energy to drive the generators. The heat is obtained by burning fuels such as oil, coal, or gas. The components of a thermal station are outlined in figure 11.22. The boiler burns the fuel and heats water to produce high-pressure steam at high-temperature. The steam is directed onto the blades of a high speed turbine which produces mechanical energy and turns the generator. The steam is condensed and the water returned to the boiler.

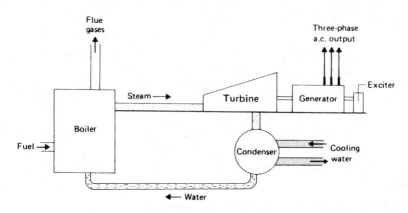

Figure 11.22 *Thermal power station scheme*

The efficiency of the boiler is the main limitation of the system but modern techniques, such as fluidised-bed combustion, offer some improvements. A thermal power station can, however, use low-quality coal and the ash recovered from the boiler can be used as a building aggregate. The condensation of the steam from the turbines requires large quantities of cooling water from a river, otherwise large and unsightly cooling towers have to be built.

The advantages of a thermal power station include the fact that they can be sited near the demand for their electricity and so save on the cost of transmission lines. It is also possible for some of the waste heat from the condensation process to be used for industrial purposes or for district heating of buildings. *Combined Heat and Power* (CHP) systems can be designed for maximum overall energy efficiency rather than maximum electrical energy with the waste of heat energy.

A disadvantage of thermal power stations is the relatively high cost of their fuels and the fact that their fuel supply will eventually be limited. The life of such a station can be as short as 20 years.

Nuclear power stations

Nuclear power stations are thermal stations where the heat energy is released from a nuclear reaction rather than by burning a fossil fuel. Radioactive elements, such as uranium and plutonium, have unstable nuclei which emit neutrons. These neutrons split neighbouring atoms, thus releasing other neutrons and producing heat. This *fission* reaction is controlled and prevented from becoming a chain reaction by using moderator materials, such as carbon, which absorb neutrons without reaction.

A nuclear reactor consists of a central container holding the radioactive fuel elements which are surrounded by control rods of the moderator material. The withdrawal of some control rods starts the reaction and the insertion of extra control rods can stop the reaction. The heat generated by the reaction is carried by a coolant from the core to a heat exchanger which then produces steam for the turbines. Because the core and the coolant emit dangerous radiation they must be well protected.

The *pressurised water reactor* (PWR) is a common type of reactor in practical use. The coolant in the reactor is water which is kept at high pressure to prevent it boiling. The *advanced gas-cooled reactor* (AGR) is an earlier type which uses carbon dioxide gas as the coolant. *Breeder reactors* are reactors that produce more fuel than they consume, as a result of the nuclear reaction.

The advantages claimed for nuclear power stations are that they reduce dependence upon fossil fuels, and that their operating costs can be lower than those of thermal stations. The disadvantages of nuclear power stations centre around the harm that escaped radioactive fuels, coolants, and waste products can cause to people and their environment. The safety of nuclear reactors depends upon the reliability with which they can be built and operated under practical conditions.

Hydroelectric power stations

Hydroelectric power stations use the kinetic and potential energy of running water to drive the generators. A large quantity of water at a height is required to provide enough energy. The original source of this energy is sunshine which lifts the water by means of evaporation. A dam usually provides both the head of water and a reservoir of stored water. The water flows down penstock pipes or tunnels and imparts energy to the water turbines which turn the generators.

The advantages of hydroelectric power stations are that their energy source is free, they have a long operating life, and they need few staff. The disadvantages of hydroelectric stations include the very large capital investment on civil engineering work. Suitable sites for such stations are limited and usually need long and expensive transmission lines to transfer energy to the consumers.

Transmission systems

Electrical energy has the useful property of being easily transferred from one place to another. A transmission system, as shown in figure 11.23, consists of conducting cables and lines, stations for changing voltages and for switching, and a method of control. The energy losses in the system must be kept to a minimum.

Alternating current is used in nearly all modern power transmission systems because it is easy to change from one voltage to another, and the generators and motors involved in AC are simpler to construct than for direct current. To obtain high transmission efficiency the current needs to be kept low because the heating losses in a line increase with the square of the current ($P = I^2R$). Large currents also require thick conductors which are expensive and heavy. To transmit large amounts of power at low current there must be a high voltage ($P = IV$). Transformers are used for obtaining the necessary high voltages and wide air gaps are used to supply the high insulation that is needed at high voltage.

Transmission lines

The *conductors* for overhead transmission lines are made of aluminium with a steel core added to give strength. The transmission of three-phase supply requires three conductors, or multiples of three, such as six conductors.

Transmission towers or 'power pylons' are needed to keep the lines spaced in the air. Air is a convenient cheap insulator but higher voltages require larger airgaps in order to prevent short circuits through the air. The lines are suspended from the towers by solid insulators made of porcelain or

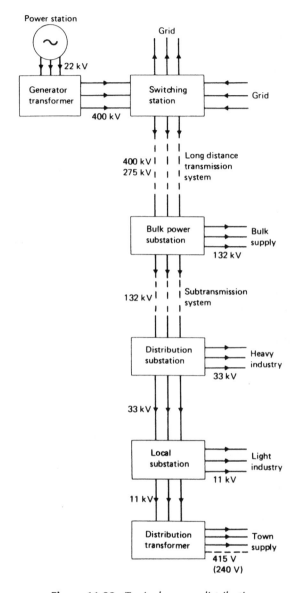

Figure 11.23 *Typical power distribution*

glass. Transmission lines operate at voltages of 132 kV and above. The United Kingdom supergrid is at 400 kV and some countries have 735 kV systems.

In a buried cable the conductor must be insulated for its whole length and protected from mechanical damage. High-voltage underground cables need special cooling in order to be efficient. It is possible to bury 400 kV transmission cables but the cost is 10 to 20 times greater than for overhead lines.

Sub-stations and switching

Power transmission systems need provision for changing voltages, for re-routing electricity and for protecting against faults.

At *sub-stations* the voltages are changed up or down as required using large transformers which are immersed in oil for the purpose of cooling as well as for insulation. At a power station, for example, a 1000 MVA transformer might step up 22 kV from the generators to 400 kV for the transmission lines. At distribution sub-stations the 400 kV is stepped down to 132 kV, 33 kV, and 11 kV.

Switching sub-stations are sub-stations where a number of transmission lines are interconnected enabling electrical energy to be routed to where it is required. When lines carrying large currents are disconnected the currents form arcs of flame which melt contacts. *Circuit breakers* are used to connect or disconnect the transmission lines and arcs are extinguished by immersion in oil or by blasts of air. Protection systems are devices which act as fuses by sensing faults on a line and then immediately isolating the line.

Transmission control

An electric power grid is a large system of interconnected power stations and consumers. The output from any power station is not dedicated to just one area but can be distributed to other areas as required.

Advantages of supply grids
* Large power stations with lower operating costs are possible
* Sudden local demands for power can be supplied by a number of power stations
* The effects of breakdowns in generating plant and transmission lines can be minimised
* Periods of low demand, such as night time, can be supplied by those plants with the lowest operating costs.

Disadvantages of grids
* It is difficult to keep all the generators working at the same frequency, which is necessary for the transfer of power
* Surplus heat from thermal power stations is often wasted.

United Kingdom electrical supply

Most of the electricity in the United Kingdom is generated by thermal power stations which make steam using gas, oil, or coal as their fuel. Some nuclear energy is used and a very small proportion of hydroelectric power is generated, mainly in Scotland.

Large thermal power stations have capacities in the range of 2000 to 4000 MW and the trend towards larger units has raised the average thermal efficiency of stations to around 35 per cent. This efficiency is measured as the ratio of the net output of electrical energy to the total input of heat energy. The total generating capacity of United Kingdom power stations is around 60 000 MW.

Typical figures for electricity consumption in the United Kingdom have shown a yearly total of 250 000 gigawatt hours (GWh). Domestic users and industry each take about 36 per cent of this total. About one fifth of the domestic electricity is used for space heating in buildings and about one sixth for water heating.

The *National Grid* is the system of interconnected transmission lines which links generating stations and consumers in England and Wales. It is one of the largest power networks under unified control in the world, although there are a variety of organisations owning different parts of the generation and distribution system.

General distribution

At suitable junctions or at the ends of transmission lines the voltage is stepped down, as indicated in figure 11.23. In the British system the voltage is reduced from 400 kV (or 275 kV) to 132 kV. The electricity is distributed at this voltage by a sub-transmission system of overhead lines to the distribution substations. At these stations the voltage is reduced to 33 kV and 11 kV for distribution by underground cable.

Large industrial consumers are supplied at 33 kV while smaller industrial consumers receive 11 kV. Small transformer stations in residential and commercial areas step the voltage down to the final 415 V three-phase, 240 V single-phase supply, which was explained in an earlier section of this chapter.

The three-phase supply is distributed by three phase cables (red, yellow, and blue) plus a neutral cable. Some commercial consumers are connected to all three cables of the supply. Households are connected to one of the phase cables and the neutral. Consumers are balanced between the three phases as evenly as possible by connecting consecutive houses to different phases in turn, for example.

Because perfect balance is not achieved the neutral cable carries a small amount of return current and is earthed at the distribution transformer. To ensure true earth potential, each consumer is supplied with an extra earth cable for attaching to the metal casing of electrical appliances. If an insulation fault causes a connection to this earth then a large current flows, immediately trips the fuse, and protects the user.

ELECTRICAL INSTALLATIONS

The electricity supply from the street is taken into a house via protected cables which connect to an electricity meter and then to the consumer unit. This unit contains an isolating switch and the means to supply the various sub-circuits in the building via suitable fuses or circuit breakers.

The system for distributing electricity within a building needs to take account of the following factors:

- Sufficient capacity for purpose
- Minimum wastage of current in the cables
- Prevention of shock
- Prevention of fire
- Means of isolation
- Compliance with regulations.

A *ring main* is used to supply power for the appliances, such as televisions, toasters or hair driers, that we connect via sockets usually on the wall. The sockets can be installed at any place on the circuit which forms a 'ring' because it is connected to the mains supply at each end. Current can be supplied to the sockets from each end of the ring main, and this arrangement allows the wiring to be of a lower current rating and smaller diameter than is required for a simple 'radial' circuit connected to the supply at one end only. To prevent a ring main being overloaded there are limits on the number of sockets and the area of building that it may serve. It is common for a house to have a separate ring main for each floor.

Within a building there is an extra conductor in the cables which does not

Table 11.1 *Features of domestic electrical installation*

Installation feature	Purpose
Main service fuse	Emergency isolation from street supply
Meter	Measurement of energy consumption
Consumer main switch	Isolation of circuits in building
Consumer control unit, bus bar	Connection for individual circuits
Circuit breakers, circit fuses	Emergency isolation of individual circuits
Earth connection	Prevention of shock
Ring main circuits, 30 amp	Power appliance
Radial circuits, 5 amp	Lighting
Radial circuit, 45 amp	Cooker, water heater
Outlet socket	Access to supply
Fuse in plug or appliance	Additional protection for appliance Prevention of shock

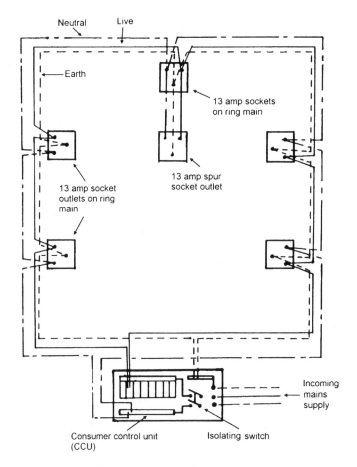

Figure 11.24 *Typical power installation*

need to exist in the street supply. This earth conductor provides a connection between the metal casing of any appliance and the ground to help protect people from electrical shock. In the case of a fault where the live supply becomes accidentally connected to the metal casing of an appliance, the *earth connection* provides a path to ground which is more efficient than any human body also touching the appliance. Large currents will instantly flow and cause the overloaded fuses or circuit breakers to react and isolate the appliance or circuit from the supply.

A more modern method of disconnecting supplies when there is danger is by the use of *residual current* devices. These use electronic circuits to monitor the current flow to and from the appliance and quickly disconnect the circuit if an incorrect flow is detected.

Exercises

1 The potential difference across a certain resistor in a television circuit is found to be 450 V when a current of 150 mA is flowing through the resistor. Calculate the value of this resistor.

2 Draw the circuit diagram of a 3 Ω and a 4 Ω resistor connected in series together, with a potential difference of 14 V applied across them.
   (a) Calculate the total current flowing in the circuit.
   (b) Calculate the voltage drop across the 3 Ω resistor.

3 Draw the circuit diagram of a 5 Ω and a 20 Ω resistor connected together in parallel, with a potential difference of 80 V applied across them.
   (a) Calculate the total resistance of the circuit.
   (b) Calculate the total current flowing in the circuit.
   (c) Calculate the current flowing in the 5 Ω resistor.

4 A 20 m length of cable carries a continuous current of 10 A. At this current the cable has a resistance of 4 mΩ/m.
   (a) Calculate the total power loss in this cable.
   (b) Calculate the total energy lost in the cable during 24 hours.

5 Calculate the current which flows in a 60 W light bulb when it is connected to a 240 V supply. How many such bulbs, wired in parallel, could be connected to a socket which is fitted with a 3 amp fuse?

6 A 136 litre hot water storage cylinder is heated by a 240 V electric element which has a resistance of 15 Ω. Assume that no energy is lost in the heating process.
   (a) Calculate the power rating of the element.
   (b) Calculate the energy needed to raise the temperature of the entire contents from 5 °C to 60 °C.
   (c) Calculate the time taken to raise the temperature as above.
      Given: density of water is 1000 kg/m^3 and specific heat capacity of water is 4200 J/kg °C.

7 A transformer with 200 turns in the primary winding is to step up voltage from 12 V to 240 V. Assume that the transformer is 100 per cent.
   (a) Calculate the number of turns needed in the secondary winding.
   (b) Calculate the current flowing in the primary winding when a 100 W lamp is connected to the output.

8 The apparent power rating of an AC motor is 4000 VA and it has a power factor of 0.85.

(a) Calculate the output power of the motor.

(b) Calculate the current drawn from the 240 V mains.

(c) Calculate the peak value of this current.

Answers

1 3 kΩ

2 (a) 2 A; (b) 6 V

3 (a) 4 Ω; (b) 20 A; (c) 16 A

4 (a) 8 W; (b) 691.2 kJ or 0.19 kWh

5 0.25 A, 12 bulbs

6 (a) 3840 W; (b) 31.42 MJ; (c) 136.4 min

7 (a) 4000 turns; (b) 8.33 A

8 (a) 3400 W; (b) 16.67 A; (c) 23.57 A

12 Water Supplies

Water is essential for animal life on Earth and is also necessary for most of our agricultural and industrial activities. A supply of good water is therefore a fundamental service for a community and this supply requires major financial and engineering investments in the systems needed for the collection, storage, treatment, and distribution of the water. The pumping of water necessary in a supply system also requires significant amounts of energy.

This chapter introduces some of the properties of liquids and uses these properties as a basis for analysing the flow of water in pipes and drains. The sources and the characteristics of natural water are described along with methods of water treatment. The features of household water supply installations are summarised.

FLUIDS AT REST

A fluid is a material whose particles are free to move their positions. Liquids and gases are both fluids and share common properties as fluids, although liquids and gases are also classed as different states of matter. The sections in this book on fluid properties are directed towards understanding the flow of water in pipes but it is useful to remember that the general principles described also explain other effects, such as the flow of air in ventilation ducts.

Pressure

The pressure on any surface is defined as the force acting at right angles on that surface divided by the area of the surface. An area which is submerged in a fluid, such as the base of a tank of water, experiences a pressure caused by the weight (which is a force) of water acting on the area of the base.

For example, The force that acts on an area A submerged at depth h in a fluid is equal to the weight of the column of liquid or gas above that area. The pressure is then this weight divided by the area.

Know   Force = weight
$\qquad$ = mass × gravitational acceleration
$\qquad$ = mg

Also know

$\quad$ Mass = density × volume
$\qquad$ = ϱV

Also know

$\quad$ Volume = Ah

Using

$$Pressure = \frac{Force}{Area}$$

$$\frac{mg}{A} = \frac{\varrho Vg}{A} = \frac{\varrho Ahg}{A}$$

The area A in the expression cancels so that, in general, the pressure at any point in a fluid is given by the following formula:

$$\boxed{p = \varrho gh}$$

where p = pressure at a point in a fluid (Pa)
$\quad \varrho$ = density of the fluid (kg/m^3)
$\quad g$ = gravitational acceleration = 9.81 m/s^2
$\quad h$ = vertical depth from surface of fluid to the point (m).

Units of pressure
- **The *pascal (Pa)* is the SI unit of pressure**

Where, by definition, 1 pascal = 1 newton/square metre (N/m^2).

Because the values of density and gravitational acceleration in the pressure equation are usually constant it is a common working practice to quote only the height associated with the pressure.

- Pressure 'head' in metres is a practical unit of pressure
- Specific weight (w) of a fluid is the product of density and gravitational acceleration, where $w = \varrho g$.

Principles of fluid pressure

- Pressure at any given depth is equal in all directions
- Pressure always acts at right angles to the containing surfaces
- Pressure is the same at points of equal depth, irrespective of volume or shape.

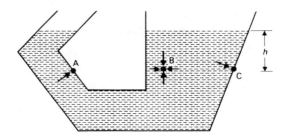

Figure 12.1 *Principles of pressure*

In figure 12.1, for example, the pressure is the same at all three points A, B and C, and is not affected by the irregular shapes involved.

Pressure measurement

In addition to the various units of pressure, described above, it is sometimes necessary to specify the base pressure being used as a reference.

- **Absolute pressure** is the value of a particular pressure measured in units above an absolute *zero* of pressure
- **Gauge pressure** is the value of a particular pressure measured in units above or below the *atmospheric* pressure.

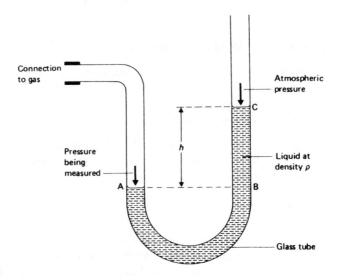

Figure 12.2 *Manometer*

For example, the pressure of the air in a car tyre is quoted as a gauge pressure and needs to be added to the value of atmospheric pressure to give absolute pressure.

Manometer

The manometer is an instrument that measures pressure by comparing levels in a U-tube containing liquid, as shown in figure 12.2. Water is a convenient liquid; oils may be used to measure some gases, and mercury is used for measuring high pressures.

In figure 12.2 the pressure at point A must be the same as the pressure at point B. The pressure at point B is produced by a column of liquid, height h, plus atmospheric pressure. This relationship can be written as follows:

Measured pressure = Atmospheric pressure + $\rho g h$

Bourdon gauge

The Bourdon gauge is an instrument that measures pressure by means of the changing curvature of a bronze tube, as shown in figure 12.3. When the pressure increases the tube tends to straighten and, by means of a mechanical linkage, this force moves a needle on a scale. The gauge needs calibrating against some absolute form of measurement.

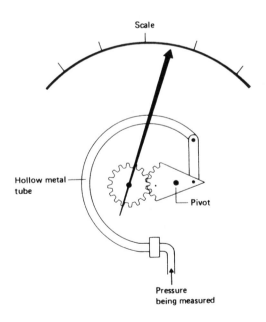

Figure 12.3 *Bourdon gauge*

Force on immersed surfaces

There is sizeable fluid pressure on a structure such as a dam or the side of a water tank and this pressure acts at right angles to the surfaces. For efficient structural design, it is necessary to know the size of the force due to the fluid pressure and any turning effect (moment) produced by the force. These same considerations apply to the design of earth-retaining walls as most soils tend to act like a fluid.

The force on such a surface can be calculated from the pressure on the surface and the area of the surface. A submerged plate is subject to different pressures and forces acting at many different points. The total force or thrust on a submerged plate is given by the following expression:

$$F = p_c A$$

where F = total resultant force or thrust (newtons N)
   p_c = pressure at the centre of area (centroid) of the immersed area (Pa)
   A = total immersed area of the plate (m²).

For a rectangular area, as shown in figure 12.4, the centre of area is positioned at half the immersed depth. Although the centre of area (centroid) is used to calculate the value of the resultant force, this force is exerted at a different position called the centre of pressure.

- ***The centre of pressure* is the point where the line of action of the resultant force passes**

The centre of pressure is always located below the centre of area of the figure. For a rectangular surface, the centre of pressure is at 2/3 of the immersed depth, measured from the top.

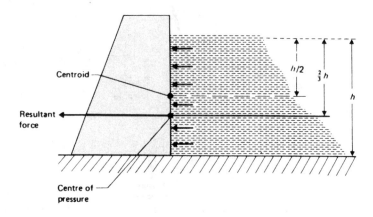

Figure 12.4 *Force on an immersed surface*

Summary

- The *size* of the resultant force is calculated using the centroid – such as the midpoint of rectangular surface.
- The *position* of the resultant force is below the centroid – such as at 2/3 of the immersed depth.

Worked example 12.1

A rectangular sluice gate is 1.6 m wide and is retaining water to a depth of 800 mm. Calculate the sideways thrust on the gate produced by the water. Given: density of water is 1000 kg/m³, and gravitational acceleration is 9.81 m/s².

Know

$\varrho = 1000\,\text{kg/m}^3$, $g = 9.81\,\text{m/s}^2$, $A = 1.6 \times 0.8 = 1.28\,\text{m}^2$
h = depth to centre of area = 0.4 m.

Using

$p_c = \varrho g h$
$\quad = 1000 \times 9.81 \times 0.4 = 3924\,\text{Pa}$

Using

$F = p_c \times A$
$\quad = 3924 \times 1.28 = 5023$

So force on gate = **5023 N**

FLUID FLOW

The behaviour of moving fluids is complex and not always fully understood. However, the construction of water supplies, drainage, and ventilation systems requires a knowledge of fluid flow. The theoretical study of fluids (hydrodynamics) is usually combined with experimental studies (hydraulics) to produce designs for practical situations.

One method of classifying fluid movement is to divide it into two major types: laminar flow and turbulent flow.

Laminar flow

Laminar or *streamline* flow produces orderly flow paths in the same direction. All particles move in fixed layers at a constant distance from the wall

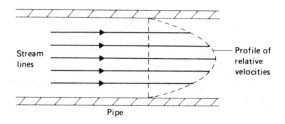

Figure 12.5 *Laminar flow*

and they do not cross one another's path. For the full pipe shown in figure 12.5 the particles at the centre have the highest velocity and some microscopic particles at the boundary have zero velocity. In general, laminar flow occurs only at low velocities.

Turbulent flow
Turbulent flow produces irregular flow paths, as shown in figure 12.6. The particles move at random, colliding with one another and exchanging momentum. Turbulent flow occurs at higher velocities and most flow in practical situations is turbulent in nature.

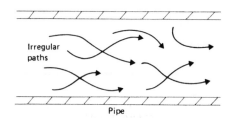

Figure 12.6 *Turbulent flow*

Transitional flow
Between the states of laminar flow and turbulent flow there is a transitional zone where the nature of the flow is complex. The change from laminar flow to turbulent flow depends upon the following factors:

- **Velocity:** increase in velocity above the *critical velocity* causes turbulent flow
- **Pipe size:** increase in pipe diameter causes turbulent flow
- **Viscosity:** decrease in viscosity ('stickiness') causes turbulent flow.

Reynolds' number

Reynolds' Number (R_e) is used to predict whether a particular flow of fluid will be laminar or turbulent in nature:

- R_e less than 2000 indicates laminar flow
- R_e greater than 2000 indicates transitional or turbulent flow.

Reynolds' number can be calculated from the following formula:

$$R_e = \frac{\varrho v D}{\mu}$$

where ϱ = density of the flowing fluid (kg/m³)

v = average velocity of fluid past a cross-section (m/s)

D = diameter of the pipe (m²)

μ = coefficient of absolute viscosity (Ns/m).

Flowrate

The amount of liquid flowing in a pipe or in a channel depends upon the dimensions of the pipe and upon the velocity of the liquid flow. This flow capacity can be described as the *discharge* or flow rate.

- **FLOWRATE (Q) is the volume of water flowing per second**

    UNIT: cubic metres per second (m³/s)

An alternative practical unit of flowrate is litres per second (l/s), where 1 m³/s = 1000 litres/s.

The definition of flowrate can also be expressed as the following formula:

$$Q = \frac{V}{t}$$

where Q = flow rate (m³/s)

V = volume of liquid passing a point (m³)

t = time taken for the volume to pass the point (s).

Consider some liquid flowing with an average velocity v in a pipe of uniform cross-sectional area A. If the pipe discharges a length l of liquid in each second then the volume of water flowing per second is derived from the following expressions:

Know

$$Q = \frac{V}{t} = \frac{lA}{t} \qquad \text{Also know} \quad v = \frac{l}{t}$$

So

$$Q = v A$$

where Q = flow rate (m³/s)
   v = average velocity of flow (m/s)
   A = cross-sectional area of pipe (m²).

Continuity equation

Figure 12.7 shows a pipe, which smoothly decreases in cross-sectional area. If the liquid flowing in the pipe is incompressible and does not vary in density then the mass of liquid flowing past each point must remain constant. The flow rate at each point in the pipe must be the same.
For continuity

$$Q_1 = Q_2$$

So

$$v_1 A_1 = v_2 A_2$$

The practical result of this continuity equation is that when the diameter of a pipe decreases then the velocity of the flow in the pipe increases.

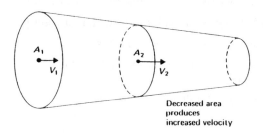

Decreased area
produces
increased velocity

Figure 12.7   *Continuity of flow*

Worked example 12.2

A storage tank measuring 2 m by 2 m is filled to a depth of 1.5 m in 5 minutes by a supply pipe with a diameter of 100 mm which runs full bore.

(a) Calculate the flow rate in the pipe.
(b) Calculate the average velocity of flow in the pipe.

(a) Know $V = 2 \times 2 \times 1.5 = 6\,\text{m}^3$, $t = 5 \times 60 = 300\,\text{s}$, $Q = ?$

Using $Q = V/t$
   $= 6/300 = 0.02$
So flow rate $= \mathbf{0.02\,m^3/s}$

(b) Know $Q = 0.02\,\text{m}^3/\text{s}$, $v = ?$

$$A = \pi \times r^2 = \pi \times \left(0.05\right)^2$$

Using $Q = vA$

$$0.02 = v \times \pi(0.05)^2$$

$$v = \frac{0.02}{\pi(0.05)^2} = 2.546$$

So velocity = **2.546 m/s**

Energy of liquids

The energy possessed by a moving liquid, is made up of the three components listed below and shown in figure 12.8:

A **Potential energy:** the energy associated with a mass of liquid at a height above a reference level (datum).
B **Pressure energy:** the energy or the work associated with moving a mass of liquid by a force.
C **Kinetic energy:** the energy associated with a mass of liquid having a velocity.

The energy components of a moving liquid are usually quantified in terms of equivalent 'heads' of liquid, measured in metres:

A: The datum head, z, is a measure of potential energy.
B: The pressure head, H, is a measure of pressure energy.
C: The velocity head, $v^2/2g$, is a measure of kinetic energy.

Bernoulli's theorem

A particle of fluid, as shown in figure 12.8, loses potential energy but gains pressure energy when it is lowered from point A to point B in figure 12.8. If the particle gains kinetic energy in moving from point B to point C then it

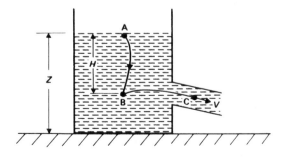

Figure 12.8 *Energy components of a liquid*

must lose some of its pressure energy or potential energy. This is a particular example of the general law of conservation of energy, which states that the total energy of a closed system must remain constant. When this principle is applied to moving fluids it is stated as *Bernoulli's theorem*:

- **BERNOULLI'S THEOREM: The total energy possessed by the particles of a moving fluid is constant**

This statement assumes that there is no loss of energy from the liquid by effects such as friction. Bernoulli's theorem can also be expressed by the following equation:

$$\frac{\text{Potential}}{\text{energy}} + \frac{\text{Pressure}}{\text{energy}} + \frac{\text{Kinetic}}{\text{energy}} = \text{Constant}$$

By referring to figure 12.9, and by writing effective pressure heads, the constancy of energy can also be written as *Bernoulli's equation*:

$$Z_1 + H_1 + \frac{v_1^2}{2g} = Z_2 + H_2 + \frac{v_2^2}{2g}$$

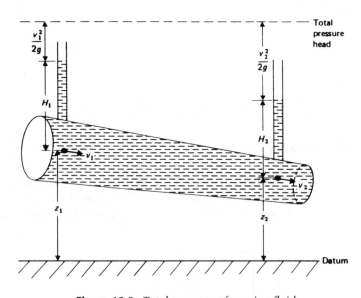

Figure 12.9 *Total pressure of moving fluid*

Pressure and velocity

A horizontal pipe has a constant datum head and constant potential energy, so that Bernoulli's theorem simplifies to the following expression:

Pressure energy + Kinetic energy = Constant

If the velocity of flowing water or air increases the kinetic energy must also increase. To keep the total energy constant the pressure energy must therefore decrease. As a general rule for a moving fluid:

- Increase in velocity gives a decrease in pressure.

This principle, derived from Bernoulli's theorem, may initially seem surprising but it explains a number of important effects associated with moving liquids and gases.

The shape of an aerofoil, such as an aircraft wing for example, causes the flowing air to have a higher velocity at the top of the wing than at the bottom. The increase in velocity produces a lower pressure on top of the wing, so that there is an upwards force on the wing.

This Bernoulli effect also causes a lifting force when strong winds blow across a pitched roof. Similarly, strong winds blowing around a building can lower pressures enough to cause windows to be 'sucked' outwards. When wind blows across the top of a chimney the pressure in the chimney decreases and causes the chimney to 'draw' smoke outwards more effectively.

Bernoulli's principle is also the basis of many devices that are used to measure the flow of liquid in rivers, channels, water mains, and sewers.

Venturimeter

The venturimeter is a device which measures the flow rate in a pipe by applying Bernoulli's principle to the pressures measured in the pipe. A constriction is constructed in the pipe and pressure gauges are fitted in the pipe and in the throat of the constriction, as shown in figure 12.10. The liquid flowing in the reduced cross-section increases in velocity and therefore, by Bernoulli's principle, it has a lower pressure. The pressure heads in the pipe and the throat are measured and used to find the flow rate.

The flow rate or discharge varies in direct proportion to the difference in pressures measured by the meter. A direct reading of flow rate can be made from a suitably calibrated gauge, or else the flow rate can be calculated from the following formula:

$$Q = C_d A a \sqrt{\frac{2g(H - h)}{A^2 - a^2}}$$

where Q = flow rate in the pipe (m³/s)
   A = cross-sectional area of pipe (m²)
   a = cross-sectional area of throat (m²)
   H = pressure head in pipe (m)

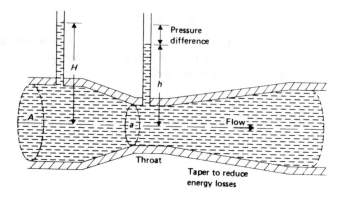

Figure 12.10 *Venturimeter*

h = pressure head in throat (m)
g = gravitational acceleration = 9.81 m/s^2
C_d = the discharge coefficient for a particular meter (usually 0.98).

Venturimeters are commonly used in water mains and the loss of energy that occurs in the meter is minimised by using a long taper downstream from the throat of the meter.

Energy losses

According to Bernoulli's theorem, the total energy of the liquid flowing in a pipe is constant. In practice there is a continuous loss of energy, even in a horizontal pipe. Friction between the liquid and the pipe wall is the main cause of energy loss and the size of the resulting pressure drop depends upon the following factors:

- Roughness of the pipe wall
- Area of the pipe wall
- Length of the pipe
- Velocity of flow
- Turbulence of flow
- Viscosity and temperature of the liquid.

The combination of these factors makes the theory of the pressure loss complicated. In order to predict the pressure losses that can be expected in a given pipeline or channel, engineers use a number of practical formulas which have been found to give reasonable results for different situations.

A general relationship that arises from theory and practice is that the pressure losses in a pipe are proportional to the square of the velocity of the liquid:

$H \propto v^2$

Re-arranging terms gives:

$$v \propto \sqrt{H}$$

The formula shows that the velocity of the flow is proportional to the square root of the pressure. So that, for example, when the pressure applied to a pipeline is doubled the velocity does not double but increases by the lesser factor of $\sqrt{2} = 1.14$.

Flow in pipes

A water mains is an example of liquid flowing in a full pipe. Energy losses through friction in the pipe cause the pressure to drop when the liquid flows, as shown in figure 12.11. To maintain the flow it is necessary to replace the energy with energy from a pump. In order to design a pumping system, the pressure losses that will occur in the pipeline need to be predicted.

Darcy's formula
Darcy's formula is one of various formulas that is used for predicting the pressure head lost from a liquid flowing in a full pipe because of friction between the liquid and the pipe surfaces. In practice, engineers may use simplified tables of friction losses for various lengths of pipe and types of fittings such as pipe elbows. The values in these table are prepared by using Darcy's formula, which can be stated as follows:

$$H = \frac{4fLv^2}{2gD}$$

where H = loss of pressure head (m)
L = length of the pipe (m)
v = average velocity of flow in pipe (m/s)
D = internal diameter of pipe (m)

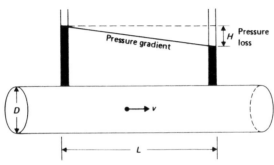

Figure 12.11 *Flow in full pipe*

g = gravitational acceleration = 9.81 m.s^2
f = Darcy's frictional coefficient (dimensionless with no units).

The value of Darcy's frictional coefficient f is found in tables and ranges from about 0.005 for smooth pipes to 0.01 for rough pipes. The value of f is also different for laminar and turbulent flow.

Flow in open channels

Water flowing in culverts, gutters and drain pipes are examples of liquid flowing in open channels or partly filled pipes. Notice that an underground drain pipe still acts like an 'open' channel for the purposes of flow. To maintain the flow in the channel the energy lost by friction is replaced by the potential energy of the liquid. The liquid must therefore always flow downhill under gravity and its total pressure head decreases as it flows.

Chézy formula

The Chézy formula expresses the relationship between the fall in height of a channel, such as a gutter, and the velocity of flow that this produces. The special definitions of gradient and radius are shown in figure 12.12. The following Chézy formula assumes that the channel has a constant gradient, and is of uniform cross-section and roughness:

$$v = c\sqrt{mi}$$

where v = average velocity of flow (m/s)
       m = hydraulic mean radius

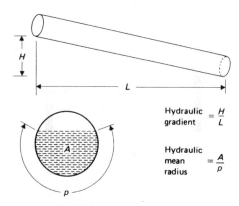

Figure 12.12 *Flow in open channel*

$$= \frac{\text{cross-section area of flow } A \ (\text{m}^2)}{\text{wetted perimeter } p \ (\text{m})}$$

i = hydraulic gradient

$$= \frac{\text{loss of pressure head } H \ (\text{m})}{\text{horizontal length of flow } L \ (\text{m})}$$

c = Chézy coefficient for a particular pipe
(for example: $c = 50 \, \text{m}^{1/2}/\text{s}$).

NATURAL WATERS

The world possesses a fixed amount of water, which is found in various natural forms, such as oceans, lakes, rivers, underground waters, ice caps, glaciers, and rain. This water plays an important part in maintaining the balance in the world's weather, especially through the presence of water vapour in the atmosphere. Water is also essential for the growth of vegetation such as trees and food crops.

Humans need a small amount of essential drinking water but much greater amounts are used for washing and waste disposal, in homes, industry, and commerce. The daily consumption of water in some cities is over 300 litres per person on average. The total supply of natural water in the Earth is enormous and should be adequate for our needs. However, local shortages do occur especially when droughts are combined with poor management of resources. This section looks at the sources of water which are used for community water supplies and describes the qualities that can be expected.

The hydrological cycle

A certain proportion of the world's natural water is involved in a continuous cycle of rainfall and evaporation. This hydrological water cycle, illustrated in figure 12.13, is made up of the following stages:

- **Evaporation:** When the Sun shines upon a surface of water, such as the ocean, some of the water evaporates from the liquid state to water vapour. The Sun also warms this water vapour which, becoming less dense than the cold air above it, rises into the atmosphere where it is circulated by weather patterns.
- **Condensation:** When the water vapour becomes cooled below its dew point it condenses and the liquid water appears as clouds of droplets.

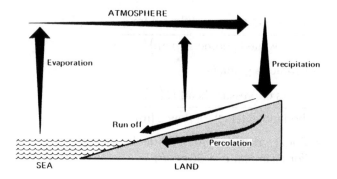

Figure 12.13 *Water cycle*

- **Precipitation:** When the water droplets in a cloud are large enough they fall as rain, snow, or hail.
- **Run off:** A proportion of the precipitated water falls on the land where most of it flows towards the sea by two main routes: as surface water such as streams and rivers; and as ground water which percolates through the materials below the surface of the land.

Plants participate in the hydrological cycle by absorbing moisture for the processes of their growth and by emitting water vapour in the process of transpiration. The energy for the hydrological cycle is provided by the Sun which supplies the latent heat for evaporation and then raises huge masses of water upwards against gravity. The many components and mechanisms of the cycle are interdependent on one another and achieve a complex natural balance which could, perhaps, be upset by major changes in some parts of the cycle.

Chemical terms

Names of compounds
Some references identify chemicals by their common names, or by outdated chemical names. Table 12.1 lists equivalent names for some of the substances which are commonly referred to in descriptions of water supplies and water treatment.

Acids and alkalis
- **An *ACID* is a substance that contains hydrogen which can be chemically replaced by other elements**

Acids have the ability to produce hydrogen ions, H^+

- **An *ALKALI* or *BASE* is a substance which neutralises an acid by accepting hydrogen ions from it**

Table 12.1 *Chemical names*

Chemical name	Chemical formula	Other names
Aluminium sulphate	$Al_2(SO_4)_3$	alum
Calcium carbonate	$CaCO_3$	chalk, limestone
Calcium hydroxide	$Ca(OH)_2$	lime (slaked, hydrated)
Calcium hydrogen carbonate	$Ca(HCO_3)_2$	calcium bicarbonate
Calcium sulphate	$CaSO_4$	gypsum, plaster
Magnesium hydrogen carbonate	$Mg(HCO_3)_2$	magnesium bicarbonate
Magnesium sulphate	$MgSO_4$	Epsom salt
Sodium carbonate	Na_2CO_3	soda, washing soda
Sodium chloride	$NaCl$	common salt, brine
Sodium hydrogen carbonate	$NaHCO_3$	sodium bicarbonate

Both acids and alkalis are corrosive and dangerous when they are very strong, but many substances are weakly acidic or alkaline when they are dissolved in water.

pH value

The pH value is a measure of the acidity or alkalinity of a solution, rated on a scale that is related to the concentration of the hydrogen ions present:

- pH less than 7 – indicates an acid
- pH greater than 7 – indicates an alkali.

The pH value of a sample can be found by chemical measurements, or by using specially calibrated electrical meters which detect the number of H^+ ions present.

Characteristics of natural water

Pure water has no colour, no taste or smell, and is neither acidic nor alkaline. Water can dissolve many substances and is sometimes termed a 'universal solvent'. It is rare for natural water to be chemically pure – even rainwater which dissolves carbon dioxide as it falls through the air. Water which flows over the surface of the land or through the ground comes into contact with many substances and takes some of these substances into solution or into suspension.

The impurities and qualities that may be found in natural waters can be conveniently described under the following headings.

Inorganic matter

Dissolved inorganic chemicals, such as salts of calcium, magnesium, and sodium cause the hardness in water, which is discussed in a later section of this chapter.

Suspended inorganic matter includes minute particles of sand and chalk, which do not dissolve in water. The particles are small enough to be evenly dispersed as a suspension which affects the colour and clarity of the water. A characteristic of a suspension is that it can be separated by settlement of the particles.

Organic matter

Dissolved organic materials usually have animal or vegetable origins and the products of their decay include ammonia compounds.

Suspended organic matter can be minute particles of vegetable or animal origin such as fibres, fungi, hair, and scales.

Micro-organisms

Diseases in man are caused by small organisms such as certain bacteria, viruses, and parasites. Some of these organisms can be carried by water if a supply is allowed to become contaminated. Examples include typhoid, cholera, and dysentery.

Pollutants

Human activities add extra impurities to natural water, mainly in the form of waste from sewage systems and from industrial processes. Domestic sewage carries disease organisms which must not be allowed to contaminate water supplies. The detergent content of household sewage can also be high and difficult to remove.

Industrial wastes which can contaminate water supplies include toxic compounds containing cyanide, lead, and mercury. Increased agricultural use of nitrogenous fertilisers, which are washed from fields or seep into the underground water sources, can lead to excessive nitrate compounds in the water supply. Certain levels of nitrates are thought to be a health hazard, especially to young children.

Acidity

Pure water is chemically pure, with a pH of 7, but natural water is invariably acidic or alkaline with a pH range of 5.5 to 8.5.

Acidity in natural water is usually caused by dissolved carbon dioxide and dissolved organic substances such as peat. Acidic waters are corrosive and also cause *plumbo-solvency* where lead, a poison, is dissolved into the water from lead tanks or pipes.

Alkalinity in natural water is more common than acidity. It is usually caused by the presence of hydrogen carbonates.

Hardness

Some natural water contains substances which form a curdy precipitate or scum with soap. No lather forms until enough soap has first been used in the reaction with the substances producing this 'hardness'.

- **HARD WATER is water in which it is difficult to obtain a lather with soap**

  UNIT: milligrammes per litre (mg/litre) of calcium carbonate ($CaCO_3$) irrespective of actual salts present

Other units: 1 part per million (ppm) is approximately 1 mg/litre; 1 'degree Clarke' is approximately 1 part per 70 000.

Typical values of hardness
   0–50 mg/litre is termed *soft* water
   100–150 mg/litre is termed *slightly hard* water
   200+ mg/litre is termed *hard* water

About 40 per cent of the public water supply in the United Kingdom is between 200 and 300 mg/litre. In general, hard water comes from underground sources or from surface water collected over ground containing soluble salts such as carbonates and sulphates; in limestone areas for example. Soft water tends to come from surface water collected over impermeable ground, such as in granite areas.

Types of hardness

There are two main types of hardness of water, defined in the following sections. The differences between types of hardness are particularly relevant for the processes of softening water which are described in a later section.

Temporary hardness
- **TEMPORARY HARDNESS is hardness that can be removed by boiling**

Temporary hardness is usually caused by the presence in the water of the following salts:

   $CaCO_3$ calcium carbonate
   $MgCO_3$ magnesium carbonate

The scale or 'fur' found inside kettles is the by-product of removing temporary hardness by boiling.

Permanent hardness

- **PERMANENT HARDNESS** is hardness in water which *cannot* be removed by boiling

Methods of removing permanent hardness are described later in the chapter. Permanent hardness is usually caused by the presence of the following salts:

$CaSO_4$ calcium sulphate
$MgSO_4$ magnesium sulphate
$CaCl_2$ calcium chloride
$MgCl_2$ magnesium chloride

Consequences of water hardness

Hardness in water has the advantages and the disadvantages listed below. Public water supplies are *not* usually treated for hardness and the suitability of an untreated supply needs to be assessed for each application of the water. Bottles of special 'mineral water' are sold at premium prices *because* they contain the minerals that cause hard water.

Disadvantages of hard water

- Wastage of fuel occurs because of scale in boilers and pipes
- Deterioration and damage to boilers and pipes is caused by scale
- Wastage of soap and energy occurs before a lather forms
- Increased wear occurs in textiles which have to be washed for longer periods
- Industrial processes are affected by the chemicals in hard water
- The preparation and final taste of food and drinks can be affected by hard water.

Advantages of hard water

- Less toxic lead is dissolved from pipes by hard water
- 'Better taste' is usually a feature of hard water
- Decreased incidence of heart disease appears to be associated with hard water.

Sources of water supply

Rainfall is the original source of the water used for drinking. Part of the water evaporates from the Earth soon after it falls as rain. Part of this water drains on the surface to join streams and rivers, and part of the water percolates into the ground to feed underground supplies.

The balance of evaporation, surface water, and underground water varies with the particular climate, the district and the time of year. A typical proportion is one third evaporation, one third run-off, and one third soak-in. A larger proportion of the rainwater is lost by evaporation during the summer.

Sources of water supply are usually classified by the routes that water has taken after rainfall. For supplies of drinking water the main categories are listed below and are described in the sections that follow:

- **Surface water** Examples include streams, rivers, lakes, and reservoirs
- **Underground water** Examples are springs and wells
- **Rainwater collectors** Examples include roofs and paved surfaces.

Underground water

When rain falls on soils or porous rocks, such as limestone or sandstone, some of the water sinks into the ground. When this water reaches a lower layer of impervious material, such as clay or rock, it may be held in a depression or it may flow along the top of the impermeable layer. Such water-bearing layers are called *aquifers*, and a cross-section of one is shown in figure 12.14. The water table or 'plane of saturation' is the natural level of the underground water. The water in some aquifers is 'confined' and held below the water table by an impermeable layer on top of the water.

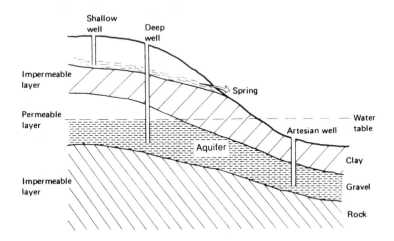

Figure 12.14 *Underground water supplies*

Springs
A spring is a source of ground water that occurs when geological conditions cause the water to emerge naturally, as shown in figure 12.14. A simple land

spring is fed by the surface water which has soaked into the sub-soil and its flow is likely to be intermittent. A main artesian spring taps water that has been flowing in an aquifer below the first impermeable layer.

The water from such a spring is usually hard with a high standard of purity achieved because of the natural purification which occurs during the percolation through the ground. All springs need protection from contamination at their point of emergence.

Wells

Wells are a source of underground water but, unlike springs, the water must be artificially tapped by boring down to the supply. Wells may be classified by the types listed below and shown in figure 12.14:

- **Shallow wells** Shallow wells tap water near the surface.
- **Deep wells** Deep wells obtain water below the level of the first impermeable layer. Such a well is not always physically deep.
- **Artesian wells** Artesian wells deliver water under their own heads of pressure, because the plane of saturation is above the ground level.

The classification of wells as 'shallow' or 'deep' depends upon the sources of water and not upon the depth of their bore. Shallow wells may give good water but there is a risk of pollution from sources such as local cesspools, leaking drains and farmyards. Deep wells usually yield hard water of high purity. The construction of all wells must include measures to prevent contamination near the surface.

Surface water

Water collected in upland areas tends to be soft and of good quality, except for possible contamination by vegetation. As a stream or river flows along its course it receives drainage from farms, roads, and towns and becomes progressively less pure. Many rivers receive sewage and industrial waste from towns and factories and are also required to supply fresh water to other towns. It is obviously important that the levels of pollution in rivers are controlled and experience has shown that even rivers flowing through areas of heavy industry can be kept clean if they are managed correctly.

A flowing river tends to purify itself, especially if the flow is brisk and shallow. This self-purification is due to a combination of factors including oxidation of impurities, sedimentation of suspended material, the action of sunlight, and dilution with cleaner water. Even so, water taken from a river for use in a large public supply usually needs treatment on a large scale and such water must be carefully analysed for its chemical and bacteriological content.

WATER TREATMENT

The variety of types and qualities of natural waters described in the previous section indicates that there is a wide range of substances whose concentrations may need to be adjusted before a water is used. The water for a public water supply is required to be 'wholesome', meaning that it is suitable for drinking.

The following properties are desirable for good drinking water:

- Harmless to health
- Colourless
- Clear
- Sparkling
- Odourless
- Pleasant tasting.

Methods of water treatment

The principal techniques used for water treatment are described in following sections and can be summarised under the general headings given below:

- **Storage:** sedimentation and clarification
- **Filtration:** slow sand filters, rapid sand filters, micro-strainers, membrane filters
- **Disinfection:** chlorination and ozonisation.

The methods used for the treatment of a particular water depend upon whether it is in small supplies or bulk supplies, and whether it is needed for domestic or industrial use. In Britain, the entire water supply is usually made suitable for drinking, even though most of it is used for non-drinking purposes.

Many industrial processes require water with less mineral content than is acceptable for drinking water and further treatment stages, such as softening, are then necessary. The addition of chemical compounds containing metals such as copper and aluminium needs to be carefully monitored and controlled.

The components of a typical water treatment works are shown in figure 12.15.

Water storage

Reservoirs are used to store reserves of water and they are also an important preliminary stage of treatment. All contaminants in the water are diluted in their effect and different qualities of water are evened out. Pathogenic (disease-producing) bacteria tend to die when in storage because of lack of suitable food, the low temperature and the action of sunlight.

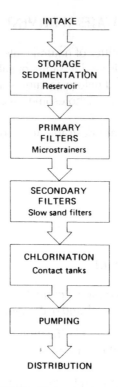

Figure 12.15 *Water treatment scheme*

A disadvantage of prolonged storage of raw water is that it provides conditions for the growth of algae: minute plants that make subsequent water treatment difficult. Algae can be controlled by the careful addition of chemicals, such as copper sulphate, and by the mixing of water in a reservoir system.

Sedimentation
Sedimentation is the gradual sinking of impurities that are suspended in the water. Simple sedimentation – the natural settling of suspended materials – takes place in reservoirs and also in specially designed settling tanks.

Clarification
Clarification is a system of chemically assisted sedimentation used for the removal of very fine suspended particles which do not settle naturally. A chemical such as aluminium sulphate (alum) produces a precipitate when it is added to the water. This precipitate coagulates with the suspended material to form a *floc*. This product of 'flocculation' then settles as a sediment, or it may be removed by mechanical collectors.

Water filtration

When water is passed through a fine material, such as sand or a wire mesh, particles are removed from the water. Some filters, such as rapid sand filters, act only as a simple physical filter and the water also requires chemical treatment. Slow sand filters, however, combine a physical action with a chemical and a bacteriological action.

Slow sand filters
Slow sand filters are built in sunken rectangular basins, with 100 m by 40 m being a typical size. A cross-section of a modern slow sand filter is shown in figure 12.16. The floor of the filter bed contains a system of collector pipes

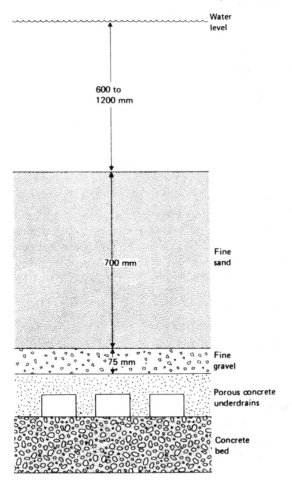

Figure 12.16 *Slow sand filter*

and underdrains covered with a layer of graded gravel. Above the gravel is a layer of sand, about 600 mm deep, which is then covered with water to a depth of around one metre.

Water slowly percolates downwards through the sand bed which develops a film of fine particles, micro-organisms, and microscopic plant life. It is this complex 'vital' layer which purifies the water by both physical and biological action. Because the growth of the active film reduces the rate of filtration the head of water is gradually increased until, after a period of weeks, the bed has to be emptied for cleaning. The top 12 to 25 mm of sand is removed, washed and eventually used for replenishing the beds. A clean filter is 'charged' by slowly filling it with water from the bottom upwards and then allowing a new vital layer to form.

The slow sand filter is extremely effective and gives high quality water, which needs little further treatment. These filters, however, occupy larger areas and work more slowly than other types of filter. The mechanical scraping and the cleaning of the sand have been made quicker and less labour-intensive by the use of machines.

Rapid gravity filters
Rapid gravity sand filters are constructed in tanks which have dimensions up to 12 m by 9 m. A uniform bed of sand, about 600 mm thick, is supported on a bed of gravel of similar thickness. Beneath the gravel is a system for collecting water and also for passing air and water upwards. The filter is filled with water to a depth of 2 to 3 m.

The water passes downwards through the sand and the filtration is mainly by physical action. The filters are cleaned by draining and then blowing a strong scour of air followed by a backwash of clean water from below. Washouts are required at intervals of 12 to 48 hours but the procedure can be automated.

Rapid gravity filters work at a rate which is some 20 to 40 times faster than slow sand filters. Their construction is more compact than slow filters and they can be cheaper to install and to operate. A rapid gravity filter is much less effective than a slow sand filter and a rapid filter is usually used in conjunction with flocculated water. Rapid filters are also used as 'primary filters' which reduce the load on the slow sand filters which follow them.

Pressure filters
Pressure filters are contained in steel pressure vessels. The construction of the sand filter inside the vessel is similar to the open rapid gravity filters and the backwashing is carried out in a similar way. The whole of the pressure cylinder is kept filled with water so that pressure is not lost during filtration and this type of filter can be inserted anywhere in a water main.

Micro-strainers

Micro-strainers are revolving drums with a very fine mesh of stainless steel wire or other material which is cleaned by water jets. The strainers are useful for screening stored waters which do not contain large amounts of suspended matter. By removing microscopic-sized particles of algae they can decrease the load on sand filters. Sometimes micro-strainers can produce water pure enough for sterilisation, without the need for filtration.

Membrane filters

Newer methods of micro-filtration or nano-filtration involve the use of membranes whose structure is so fine that they can physically sieve out bacteria and other organic pollutants such as herbicides. A membrane cartridge, for example, contains bundles of polymer fibres whose surfaces have tiny pores which are small enough to trap impurities greater than 0.1 micron (1 micron $= 10^{-6}$ metre). As the molecules of most bacteria are between 0.5 and 1.0 micron in diameter this micro-filtration is capable of providing safe drinking water. Untreated water is pumped under pressure through collections of membrane cartridges arranged in banks. Like other filters the membranes have to be cleaned by backwashing the surfaces with water and this process can be automated.

Disinfection

The disinfection of water supplies is intended to reduce harmful organisms, such as bacteria, to such very low levels that they are harmless. This safe quality needs to be maintained while the water is in the distribution system, including any reservoirs which store purified water. Disinfection can be achieved by a number of agents but chlorine and ozone are usually employed to treat public water supplies.

Chlorination

Chlorine is a powerful oxidising agent which attacks many organic compounds and has the property of killing bacterial cells. The rate of disinfection of a particular type of water depends upon the chemical form of the chlorine and upon the length of time it is in contact with the water. The presence of ammonia in the water reduces the effectiveness of the chlorine, but the addition of ammonia after disinfection can be useful for reducing possible objectionable tastes produced by the chlorine.

The chlorine is usually kept liquefied in tanks from where it is injected into the water at a controlled rate. Because the sterilising effect is not instantaneous the water and the chlorine are held in contact tanks for a period of time, such as 30 minutes.

Ozone treatment

Ozone, O_3, is a form of oxygen with molecules that contain an extra atom. This form of oxygen is unstable and reactive so ozone is a powerful oxidising agent which rapidly sterilises water. Ozone treatment is more expensive than chlorination as the ozone must be made at the treatment station. Air is dried, subjected to a high voltage electric discharge and the ozone produced is then injected into the water.

Softening of water

Hard water is satisfactory, and even desirable, for drinking but hard water also has disadvantages described previously. It is *not* usual practice to soften public supplies of water but softening may be necessary for industrial supplies. There are several different principles used for changing hard water into soft water:

- **Precipitation:** Precipitation methods act by completely removing most of the hardness compounds which are present in the water. Chemicals are added to the hard water to form insoluble precipitates which can then be removed by sedimentation and filtration.
- **Base exchange:** Base exchange methods act by changing hardness compounds into other compounds which do not cause hardness.
- **Demineralisation:** Demineralisation is the complete removal of all chemicals dissolved in the water. This can be achieved by an ion-exchange which is a more complete process than the base-exchange.

Lime–soda treatment

'Lime' and 'soda' processes are precipitation methods of water softening. They depend upon chemical reactions to make the calcium and magnesium content of the hard water become insoluble precipitates which can then be removed. The following softening reactions use the chemical terms listed in table 12.1:

Calcium temporary hardness is treated by *lime*:

$$Ca(HCO_3)_2 + \underset{\substack{[\text{slaked} \\ \text{lime}]}}{Ca(OH)_2} \rightarrow \underset{[\text{precipitate}]}{2CaCO_3} + 2H_2O$$

Calcium permanent hardness is treated by *soda*:

$$CaSO_4 + Na_2CO_3 \rightarrow CaCO_3 + Na_2SO_4$$

or

$$CaCl_2 + \underset{[\text{soda ash}]}{Na_2CO_3} \rightarrow \underset{[\text{precipitate}]}{CaCO_3} + 2NaCl$$

Magnesium temporary hardness is treated by *lime*:

$$Mg(HCO_3)_2 + \underset{\substack{[\text{slaked} \\ \text{lime}]}}{2Ca(OH)_2} \rightarrow \underset{[\text{precipitate}]}{2CaCO_3} + Mg(OH)_2 + 2H_2O$$

Magnesium permanent hardness is treated with *lime and soda*:

$$MgSO_4 + Ca(OH)_2 \rightarrow Mg(OH)_2 + CaSO_4$$

or

$$MgCl_2 + \underset{\substack{[\text{slaked} \\ \text{lime}]}}{2Ca(OH)_2} \rightarrow Mg(OH)_2 + \underset{\substack{[\text{treat as for} \\ \text{calcium}]}}{CaCl_2}$$

In a lime–soda softening plant the chemicals are measured and added to the water, either dry or in a slurry. The solids formed in the precipitation are removed by settlement in tanks, or by flocculation, and the water is then filtered. Lime–soda water softening processes have relatively low running costs but produce large quantities of sludge which require disposal.

Base-exchange method

In the base-exchange method of water softening the hardness-forming calcium and magnesium salts are converted to sodium salts which do not cause hardness. *Zeolites* are the special materials which act as a medium for the exchange of ions:

calcium   sodium   calcium   sodium
sulphate + zeolite → zeolite + sulphate
(hard)   (soft)

The exhausted zeolite is 'regenerated' with sodium from a salt solution:

calcium   sodium   sodium   calcium
zeolite   +   chloride   →   zeolite   +   chloride

Natural zeolites are obtained from processed sands and synthetic zeolites are made from organic resins such as those of polystyrene. The water softener is usually a metal cylinder, constructed like a pressure filter, in which the water passes downwards through a bed of zeolite. When regeneration is necessary dilute brine is passed through the bed, followed by a freshwater wash.

The base-exchange process can be operated simply and automatically but its costs depend upon the availability of salt. The process is used for commercial supplies such as for boilers and laundries and for household water softeners.

Scale prevention

The build-up of scale or 'fur' within boilers and pipes can be minimised by techniques which do not actually soften the hard water. Instead the crystalline growth of the scale is inhibited or interrupted so that it does not form on surfaces.

These techniques include the use of chemicals, usually phosphates, such as those which are added to the closed circuit of a hot-water heating system. Non-chemical methods include the use of mechanical vibrations, electrical and magnetic effects.

Magnetic water treatment
In magnetic water treatment the hard water flows through a 'water conditioner' which applies a strong magnetic field to the water. When this water is heated the hard water products remain as microscopic particles suspended in the water instead of forming scale. The products are then carried by the movement of the water or else form a movable sediment.

WATER INSTALLATIONS

Distribution to buildings

After water has been treated it is pumped to the start of a local distribution system which is often a high-level storage reservoir or a water tower. Water can then be supplied by gravity through iron pipes or polymer pipes (coloured blue) beneath the streets. These pipes flow full of water under pressure so secure joints are important in order to prevent leakage. Damage to pipes caused by age, ground movement or heavy traffic can cause a serious wastage of water.

Domestic water installations

The system for distributing water within a building needs to take account of the following factors:

- Sufficient capacity for purpose
- Leakproof pipework
- Means of isolating pipework appliances
- Means of draining pipework and appliances
- Arrangements for overflows
- Prevention of back pollution to the public supply
- Compliance with regulations.

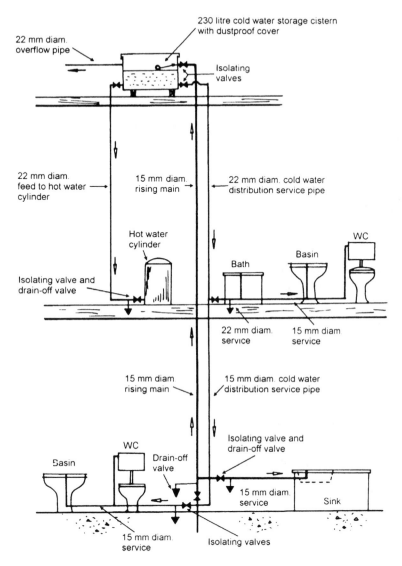

Figure 12.17 *Typical indirect cold water installation*

Once inside the building the cold water can be distributed to points inside the buildings by two main methods: *direct* systems where water is taken straight 'up' from the mains, and *indirect* systems where water is taken 'down' from storage tanks, often in the roof space.

The features of the different methods of water supply are listed below and illustrated in figure 12.17.

Table 12.2 *Features of water installation*

Installation feature	Purpose
Street valve	Emergency isolation from street supply
Meter	Measurement of water consumption
Storage cistern	Constant pressure
	Continuity of supply
	Isolation from public supply
Drain-off valve	Means of draining system
Overflow pipe	To divert overflow water outside
22 mm and 15 mm diam. service pipes	To provide appropriate capacity
Isolating valves	To service appliances

Indirect cold water supply

Features of indirect cold water supplies include the following:

- Water stored in cistern at higher level
- Mains is fed to storage cistern
- Cold taps not for drinking
- Demand from mains is smoothed
- Building is protected from mains failure
- More plumbing installation and higher costs.

Direct cold water supply

Features of direct cold water supplies include the following:

- No storage of water involved
- Mains supply is fed directly to all outlets
- All cold taps suitable for drinking
- Higher peak demands on mains
- Risk of back-syphonage to mains
- Less plumbing installation and lower costs.

Exercises

1 When an oil-filled manometer measures a certain pressure the difference in oil levels is 240 mm. Express this pressure as an absolute pressure. Given: density of the oil is 830 kg/m^3; gravitational acceleration is 9.81 m/s^2.

2 A reservoir dam, 200 m in length, has a vertical face which retains water to a depth of 15 m. Calculate the horizontal force on the dam produced

by the retained water. Given: density of water is $1000\,kg/m^3$; gravitational acceleration is $9.81\,m/s^2$.

3 A 200 mm diameter pipe runs into a 150 mm diameter pipe with a discharge rate of $0.04\,m^3/s$. If both pipes are running full bore then calculate the flow velocities in each pipe.

4 A venturi meter has a throat of 80 mm diameter and is set in a horizontal water main of 150 mm diameter. If the measured pressure heads are 13.7 m in the main and 10.5 m in the throat then calculate the flow rate in the pipe. Given: discharge coefficient for the meter is 0.98; gravitational acceleration is $9.81\,m/s^2$.

5 A 100 mm diameter water main is required to discharge at $0.035\,m^3/s$ when running full bore. Calculate the loss of pressure head that occurs in a 150 m horizontal run of this pipe. Use Darcy's formula and assume a friction coefficient of 0.006 for the pipe.

6 A circular drain of 150 mm diameter is laid with a fall of 1 in 40 and is running half full of water. Assume a Chézy coefficient is $50\,m^{1/2}/s$.
 (a) Calculate the flow velocity in the drain.
 (b) Calculate the discharge rate (in m^3/s) for these conditions.

Answers

1 1954 Pa

2 220.7 MN

3 1.273 m/s, 2.264 m/s

4 $0.041\,m^3/s$

5 36.43 m

6 **(a)** 1.531 m/s; **(b)** $0.0135\,m^3/s$

13 Environmental Buildings

This chapter reviews some interactions between buildings and our wider environment. The correct assessment of climate helps to create buildings which are successful in their external environment, while a knowledge of 'sick buildings' helps to avoid unsuccessful internal environments. The sections on energy conservation and green buildings suggest how the correct design and use of our buildings can help to improve our total environment.

CLIMATE

A fundamental reason for the existence of a building is to provide shelter from the climate, such as the cold and the heat, the wind and the rain. The climate for a building is the set of environmental conditions which surround a building and links to the inside of a building by means of heat transfer.

Climate has important effects on the energy performance of buildings, in both winter and summer, and on the durability of the building fabric. Climates which are favourable to energy use and durability also make the external environment of a building attractive and useful for recreation.

Although the overall features of the climate are beyond our control, the design of a building can have a significant influence on the climatic behaviour of the building. The following measures can be used to enhance the interaction between buildings and climate:

- Selection of site to avoid heights and hollows
- Orientation of buildings to maximise or minimise solar gains
- Spacing of buildings to avoid unwanted wind and shade effects
- Design of windows to allow maximum daylight in buildings
- Design of shade and windows to prevent solar overheating
- Selection of trees and walls surfaces to shelter buildings from driving rain and snow
- Selection of ground surfaces for dryness.

Climate types

The large-scale climate of the Earth consists of interlinked physical systems powered by the energy of the Sun. The built environment generally involves the study of smaller systems for which the following terms are used:

- **Macroclimate:** The climate of a larger area, such as a region or a country
- **Microclimate:** The climate around a building and upon its surfaces.

A building site may have natural microclimates caused by the presence of hills, valleys, slopes, streams, and other features. Buildings themselves create further microclimates by shading the ground, by drying the ground, and by disrupting the flow of wind. Further microclimates occur in different parts of the same building, such as parapets and corners, which receive unequal exposures to sun, wind and rain.

Effects of microclimate

An improved microclimate around a building brings the following types of benefits:

- Lower heating costs in winter
- Reduction of overheating in summertime
- Longer life for building materials
- Pleasant outdoor recreation areas
- Better growth for plants and trees
- Increased user satisfaction and value.

Climatic data

In order to design a building which is appropriate for its site, the climate of that site needs to be studied and predicted. The following climatic factors can be considered:

- Temperatures
- Humidity
- Precipitation of rain and snow
- Wind speed and direction
- Sunshine hours and solar radiation
- Atmospheric pollution.

These factors can vary by the hour, by the day, and by the season. Some of the variations will cycle in a predictable manner like the Sun, but others such as wind and cloud cover will be less predictable in the short term. Information about aspects of climatic factors is collected over time and made available in a variety of data forms including the following:

- Maximum or minimum values
- Average values
- Probabilities or frequencies.

The type of climatic data that is chosen depends upon design requirements. Peak values of maximum or minimum are needed for some purposes, such as sizing heating plant or designing wind loads. Longer term averages, such as seasonal information, are needed for prediction of energy consumption. Some measurements in common use are described in the following sections.

Degree-days, Accumulated Temperature Difference

The method of degree days or Accumulated Temperature Difference, ATD, is based on the fact that the indoor temperature of an unheated building is, on average, higher than the outdoor. For a traditional British construction the difference is taken to be 3 °C.

In order to maintain an internal design temperature of 18.5 °C, for example, the building only needs heating when the outdoor temperature falls below 15.5 °C (18.5 − 3). This *base temperature* is used as a reference for counting the degrees of outside temperature drop and the number of days for which such a drop occurs.

- One day at 1 °C below base temperature gives 1 degree-day
- Two days at 1 °C below base temperature gives 2 degree-days, or
- One day at 2 °C below base temperature also gives 2 degree-days.

The accumulated temperature difference total (degree-days) for a locality is a measure of climatic severity during a particular season and typical

Table 13.1 *Climatic severity*

Area	Degree-days
England	
South West	1800–2000
South East	2000–2100
Midlands	2200–2400
North	2300–2500
Wales	2000–2200
Scotland	2400–2600

Notes
Accumulated Temperature Differences using base temperature of 15.5 °C, September to May.
Increase Degree-day values by 200 for each 100 m above sea level.

values are given in table 13.1. This data, averaged over the years, can be used in the calculation of heat loss and energy consumption. ATD totals do not take account of extra heat losses caused by exposure to wind and of heat gains from solar radiation.

Driving rain index, DRI

The annual driving rain index, DRI, is a combined measure rainfall and wind speed. The DRI takes account of the fact that rain does not always fall vertically upon a building and that rain can therefore penetrate walls. The DRI is also associated with the moisture content of exposed masonry walls whose thermal properties, such as insulation, vary with moisture content. Damp walls have poorer insulation than dry walls.

Driving rain is usually caused by storms but intense driving rain can also occur in heavy showers which last for minutes rather than hours. These conditions are more likely in exposed areas, such as coasts, where high rainfall is accompanied by high winds. Table 13.2 gives typical values of driving rain index for different types of area.

Maps are published which enable the driving rain index to be predicted for different parts of Britain. Variations in microclimates and in types of building make an exact correlation difficult but, in general, it has been predicted that in a 'sheltered' region a one-brick thick solid wall would not suffer from rain penetration. In addition to damp walls, rain penetration through poorly-sealed windows often causes problems in areas of severe exposure.

Table 13.2 *Driving-rain indices for British Isles*

Exposure grading	Driving-rain index	Example
Sheltered	3 or less	Within towns
Moderate	3 to 7	Countryside
Severe	7 or more	West coastal areas

Note: High buildings, or buildings of any height on a hill, usually have an exposure one degree more than indicated.

Wind data

The main effects of wind on a building are those of force, heat loss, and rain penetration. These factors need to be considered in the structural design and in the choice of building materials. *Wind chill* factor relates wind to the rate of heat loss from the human body rather than the loss from buildings. The unfavourable working conditions caused by wind chill have particular relevance to operations on exposed construction sites and tall buildings.

Wind speed

The force of a wind increases with the square of the velocity, so that a relatively small increase in wind speed produces a larger than expected force on a surface such as a building. The cooling effect of wind, measured by wind chill, also increases greatly with the speed of the wind. Typical wind speeds range between 0 m/s and 25 m/s, as described below:

- 5 m/s wind disturbs hair and clothing
- 10 m/s wind force felt on body
- 15 m/s wind causes difficulty walking
- 20 m/s wind blows people over.

The air flow around some parts of a building, especially over a pitched roof, may increase sufficiently to provide an aerodynamic lifting force by using the principle of Bernoulli described in chapter 12. This force can be strong enough to lift roofs and also to pull out windows on the downwind side of buildings.

Wind direction

The direction of the wind on a building affects both the structural design and the thermal design. The directional data of wind can be diagrammatically shown by a 'rose' of arms around a point, to represent the frequency that the wind blows from each direction.

The length of each arm of the rose can be proportional to the number of days that the wind blows from the direction shown by that particular arm. Directions with longer arms will indicate colder winds which affect energy consumption. Other systems of roses may indicate the direction of wind chill factors which affect human comfort and operations on a building site.

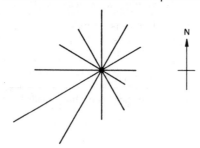

Figure 13.1 *Directional wind rose of mean wind speeds*

Wind effects around buildings

The existence of buildings can produce unpleasantly high winds at ground level. It is possible to estimate the ratio of these artificial wind speeds to the wind speed that would exist without the building present. A typical value of *wind speed ratio* around low buildings is 0.5, while around tall buildings the ratio might be as high as 2. A wind speed ratio of 2 will double normal wind speed.

- A maximum wind speed of 5 m/s is a suitable design figure for wind around buildings at pedestrian level.

General rules for the reduction of wind effects are given below:

- Reduce the dimensions, especially the height and the dimensions facing the prevailing wind
- Avoid cubical large shapes
- Use pitched roofs rather than flat roofs; use hips rather than gable ends
- Avoid parallel rows of buildings
- Avoid funnel-like gaps between buildings
- Use trees, mounds and other landscape features to provide shelter.

Solar data

The effects of the Sun on buildings is considered under the topics of Heat Gains in chapter 3 and Natural Lighting in chapter 7. The prediction of these effects requires the following categories of knowledge about the Sun:

- Position in the sky and the angle made with building surfaces
- Quantity of radiant energy received upon the ground or other surface
- Obstructions and reflections caused by clouds, landscape features and buildings.

Sun position

The path that the Sun makes across the sky changes each day but repeats in a predictable manner which has been recorded for centuries. For any position of the Sun, the angle that the solar radiation makes with the wall or roof of a building can be predicted by geometry. This angle of incidence has a large effect as the energy received as solar radiation obeys the *Cosine Law of Illumination*, described in chapter 5, so that intensity is at a maximum when the radiation strikes a surface at right angles.

Figure 13.2 is a simple form of sunpath diagram showing the position of the Sun in the sky at different times of the year. Other graphical forms of sunpath diagram and 'sky maps' allow the dimensions and orientation of buildings and landscape to be plotted on the same diagram in order to predict the radiation on the building surfaces.

Another design approach has been to use a model of the building and a movable light source, known as a *helidon*, which can imitate the movements of the Sun. Traditional methods of solar prediction are being replaced by *computer* methods for assessing the directions and quantities of solar radiation falling upon a building. In addition to numerical results,

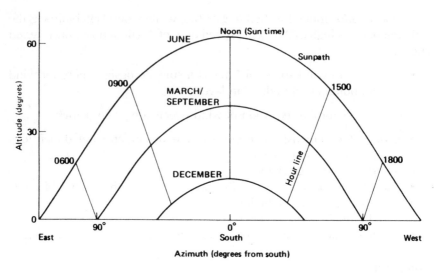

Figure 13.2 *Sunpaths for Southern England*

these programs can produce pictures of the effects of sunshine and shade upon a building, or upon a complete building site, at any time of the day or year.

Solar radiation

The intensity of solar radiation falling on a surface, such as the ground, can be measured in watts per square metre (W/m^2) of that surface. The watt is defined as a joule per second so this is an instantaneous measurement of the energy received per second on each square metre. When the solar energy is measured over a period of time, such as a day or year, the units will be joules or megajoules per square metre (MJ/m^2). A rounded figure for the intensity of solar radiation available on many parts of the Earth's surface is $1000\,W/m^2$. This is a valuable potential source of energy although only a fraction of it can usually be converted for use.

For Southern England, figure 3.5 in the section on Heat Gains, shows a June peak value of $850\,W/m^2$ on a horizontal surface. When clouds obscure direct sunshine, significant amounts of diffuse radiation are still available. Annual *totals* of received solar radiation only vary by a factor of about 1.0 and 1.3 between the south and the north of the United Kingdom.

ENERGY CONSERVATION

At present, most of the energy used to heat buildings, including electrical energy, comes from fossil fuels such as oil and coal. This energy originally came from the Sun and was used in the growth of plants such as trees. Then, because of changes in the Earth's geology, those ancient forests eventually became a coal seam, an oil field, or a natural gas field. The existing stocks of fossil fuels on Earth cannot be replaced and, unless conserved, they will eventually run out.

In the United Kingdom, some 40 to 50 per cent of the national consumption of primary energy is used for building services such as heating, lighting, and electricity. Over half of this energy is consumed in the domestic sector and most of that energy is used for space heating. Reducing the use of energy in buildings will therefore be of great help in conserving energy resources and in saving money for the occupants of buildings.

Methods of conserving energy in buildings are influenced by the costs involved and, in turn, these costs vary with the types of building and the current economic conditions. Some of the important options for energy conservation in buildings are outlined below; articles in newspapers and journals are also helpful for staying informed about the balance and interactions of these matters.

Alternative energy sources

It is difficult to do without oil and coal for certain purposes, such as converting them to chemicals and making polymer products like plastics and paints. It is not essential, however, to use fossil fuels for the heating of buildings. There are alternative energy sources available for buildings and the energy consumption of buildings can also be greatly reduced by changes in their design and use.

Solar energy
All buildings gain some casual heat from the Sun during winter but more use can be made of solar energy by the design of the building and its services. Despite the high latitude and variable weather of countries in North-Western Europe, like the United Kingdom, there is considerable scope for using solar energy to reduce the energy demands of buildings.

The utilisation of solar energy need not depend upon the use of special 'active' equipment such as heat pumps. *Passive solar design* is a general technique which makes use of the conventional elements of a building to perform the collection, storage, and distribution of solar energy. For example, the afternoon heat in a glass conservatory attached to a house can be stored by the thermal capacity of concrete or brick walls and floors. When this heat is given off in the cool of the evening it can be circulated into the house by natural convection of the air.

Natural sources
Large amounts of energy are contained in the world's weather system, which is driven by the Sun, in the oceans, and in heat from the Earth's interior

which is caused by radioactivity in rocks. This energy is widely available at no cost except for the installation and running of conversion equipment. Devices in use include electricity generators driven by windmills, wave motion, and geothermal steam.

Energy efficiency

The total energy of the Universe always remains constant but when we convert energy from one form to another some of the energy is effectively lost to use by the conversion process. For example, hot gases must be allowed to go up the chimney flue when a boiler converts the chemical energy stored in a fuel into heat energy. Around 90 per cent of the electrical energy used by a traditional light bulb is wasted as heat rather than light.

Efficient equipment

New techniques are being used to improve the conversion efficiency of devices used for services within buildings. Condensing boilers, for example, recover much of the latent heat from flue gases before they are released. More efficient forms of electric lamp have been described in the chapter on Lamps and Luminaires. Heat pumps can make use of low temperature heat sources, such as waste air, which have been ignored in the past.

Electricity use

Although electrical appliances have a high energy efficiency at the point of use, the overall efficiency of the electrical system is greatly reduced by the energy inefficiency of large power stations built at remote locations. The 'cooling towers' of these stations are actually designed to waste large amounts of heat energy.

It is possible to make use of this waste heat from power stations for various uses in industry and for the heating of buildings. These techniques of CHP (Combined Heat and Power) can raise the energy efficiency of electricity generation from around 33 per cent to as much as 80 per cent. CHP techniques can also be applied on a small scale to meet the energy needs of just one building or a series of buildings.

In the absence of CHP, savings in national energy resources are made if space heating is supplied by burning fossil fuels, such as coal or gas, inside a building that is to be heated rather than by burning those fossil fuels at a power station. Electrical energy will still be required for devices such as lights, motors, and electronics but need not be used for heating.

Thermal insulation

External walls, windows, roof and floors are the largest areas of heat loss from a building and are discussed under Thermal Insulation in chapter 2.

Current standards of thermal insulation in the United Kingdom, as defined by Building Regulations, have scope for continued improvement. For example, the practice of 'trading off' a well-insulated area of a building against a poorly-insulated area makes it possible to use areas of inferior insulation.

The upgrading of insulation in existing buildings can be achieved by techniques of roof insulation, cavity fill, double-glazing, internal wall-lining, and exterior wall-cladding. Increasing the insulation without disrupting the look of a building is a particular challenge for countries with stocks of traditional buildings.

Ventilation

The warm air released from a building contains valuable heat energy, even if the air is considered 'stale' for ventilation purposes. The heat lost during the opening of doors or windows becomes a significant area of energy conservation, especially when the cladding of buildings is insulated to high standards.

These ventilation losses are reduced by better seals in the construction of the buildings, by air-sealed door lobbies, and the use of controlled ventilation. Some of the heat contained in exhausted air can be recovered by heat exchange techniques such as heat pumps.

SICK BUILDINGS

There are some buildings, especially offices, where the occupants appear to suffer ill health more often than might reasonably be expected. These 'sick building' illnesses have no readily identifiable cause. Some illnesses, such as Legionnaires' disease, which can be traced to a particular cause are *not* strictly sick building illnesses.

Sick building syndrome (SBS) is the most commonly-used term for the phenomenon and is recognised by the World Health Authority. Other terms found in use include: Building Related Illness (BRI); Tight Building Syndrome (TBS); office eye syndrome and others.

It has been suggested that up to 30 per cent of new and refurbished buildings have given rise to complaints of sick building illness. Although this consideration of sick building syndrome is centred around office environments, it is well to consider that the home environment can be troubled by many of the same physical causes, if not the mental causes.

Sick building effects

The illnesses related to sick building syndrome generate the following types of symptoms:

- Eye, nose and throat irritations
- Dryness of throat, nose and skin
- Breathing difficulties and chest tightness
- Headaches, nausea, dizziness
- Mental fatigue
- Skin rashes
- Aching muscles and 'flu-like symptoms.

This wide range of symptoms includes illnesses which most people occasionally suffer while at home or at work so investigations are difficult. However, certain patterns have been found.

Sick building syndrome is linked to the size and structure of the organisation using the office. Most complaints occur in offices which contain many staff and symptoms are more frequent in the afternoon than the morning. The people with most symptoms are those who see themselves having least control over their environment and clerical staff are more likely than managerial staff to suffer from SBS.

Whatever the causes, a sick building results in absenteeism among staff and lower productivity while they are at work. Sick building syndrome also costs money by loss of profit, by bad publicity and, in the extreme, by closure of a building.

Buildings at risk

Despite the difficulties of investigating sick buildings syndrome, the following features have been identified as common to many sick buildings:

- Forced ventilation, including mechanical ventilation and air conditioning
- Windows and other openings sealed for energy efficiency
- Lightweight construction
- Carpets and other textiles used on indoor surfaces
- Warm and uniform environments.

Sick building causes

The following list of factors which contribute to sick building syndrome have been arranged in three general headings of physical, chemical, and microbial.

Physical comfort conditions
- Uncomfortable temperatures
- Low humidity
- Low air movement and 'stuffiness'
- Low ventilation rates

- Insufficient negative air ions
- Unsuitable lighting and decoration
- Low daylight levels
- Uncomfortable seating
- Excessive noise levels
- Electromagnetic radiation from electrical services and appliances
- Low morale and general dissatisfaction.

Chemical pollutants
- Cigarette smoke
- Formaldehyde vapours from furniture, particle boards
- Organic vapours from adhesives, paints, and cleaners
- Radon decay products from granite stone and aggregates
- Ozone gas from photocopiers, laser printers and high-voltage sources.

Microbial
- Airborne micro-organisms from bacteria and fungi in air conditioning systems
- Micro-organisms in drinking water and vending machines
- Micro-organisms in carpets, fabrics, and pot plants.

Sick building solutions

The previous list of possible causes of sick building illnesses highlight the fact that is not possible to identify single factors and, therefore, there is no single cure. Of the physical causes, poor air quality and dirty machinery are common. The whole problem of sick buildings centres around human beings and the study of the people concerned is as important as the surroundings.

The following general guidelines will help to eliminate the various causes which give rise to poor environments within buildings:

- Good design
- Correct installation
- Constant maintenance.

These measures must be directed towards creating a healthy and pleasant working environment for the occupants of a building and these occupants need to feel involved in the creation and control of their environment.

GREEN BUILDINGS

The processes of constructing and running a building can have a major effect upon the environment. Effects on the local area, such as the visual impact of a building, have always been recognised but now it is also realised that buildings contribute to larger effects, such as global warming and wastage of resources.

The energy used in buildings is responsible for around half of the total production of the carbon dioxide gas which is the principal cause of the greenhouse effect in the atmosphere. The CFCs used in some building insulation materials and in refrigerants also contribute to the depletion of the ozone layer.

In the United Kingdom, the Building Research Establishment supervises a system of rating buildings for their environmental friendliness. The Building Research Establishment Environmental Assessment (*BREEAM*) system appraises buildings under the following criteria.

Global issues
- Global warming, linked to the burning of fossil fuels
- Ozone depletion, linked with the use of CFCs in insulants and coolants
- Rain forest destruction, linked with the use of hardwood from forests which are not renewed
- Resource depletion, linked with non-recycling of materials.

Neighbourhood issues
- Legionnaires' disease, linked with faulty air conditioning systems
- Local wind effects, linked with the height and shape of buildings
- Re-use of a site previously built upon or reclaimed.

Indoor effects
- Legionnaires' disease, linked with hot and cold water supplies
- Lighting, such as modern fluorescent lamps with electronic control gear
- Indoor air quality linked with ventilation rates, humidification, control of systems and separate areas for smoking
- Hazardous materials, such as asbestos, formaldehyde in cavity fill, and lead in paint.

The list of items considered in the environmental assessment of a building will change as knowledge of risks associated with materials and techniques evolves.

INTELLIGENT BUILDINGS

When a building is described as 'intelligent' it is considered to have a high proportion of the following features:

- **Automated building services** such as those for energy management, security and fire precautions.
- **Information management** such as telecommunication systems and computer systems for IT (Information Technology).
- **Connectivity** determined by internal cabling and access to external services.
- **Control** of the environment achieved by the monitoring and control of building services and safety such as in a *building automation system* (BAS).
- **Premises management** achieved by the controlled monitoring and scheduling of maintenance and other building functions.

Information technology

The attributes of intelligent buildings, described above, depend largely on the use of electronics, microcomputers, and associated software. Information Technology (IT) therefore plays an important role in the choice of buildings in addition to being involved in their design and construction processes.

INTEGRATED BUILDING DESIGN

Although the various topics that make up environmental technology are usually defined and studied separately, it is important to appreciate their dependence upon one another when they are combined together in a building. All of the environmental factors have ideal standards to be separately satisfied but they may also need to be balanced against one another. As a simple example, larger windows provide better daylighting but they also cause greater heat losses in winter and larger heat gains in summer. The environmental factors listed in table 13.3 indicate some of the major interactions between different design decisions.

The topics studied as environmental science must be considered together and at an early stage in the design. The need for this integrated and early involvement of environmental studies has often been neglected. Historically, engineers have specialised in one particular aspect of controlling the environment of a building and have not considered the effects of their decision on other areas of the environment.

As a result, many modern buildings are environmental failures, despite the improved technology available. In order to provide the best possible environment in future buildings all available knowledge and skill need to be

Table 13.3 *Interactions of environmental decisions*

Some design options	Possible environmental effects			
	Heating	Ventilation	Lighting	Sound
Sheltered site loss and gain	Less heat	–	Less daylight	Less noise intrusion
Deep building shape	Less heat loss and gain	Reduced natural ventilation	Less daylight	–
Narrow building plan	More heat loss and gain	More natural ventilation	More daylight intrusion	More noise
Heavy building materials	Slower heating and cooling	–	–	Better sound insulation
Increased window area	More heat loss and gain	–	More daylight	More noise intrusion
Smaller, sealed windows	Less heat loss and gain	Reduced natural ventilation	Less daylight	Less noise intrusion

integrated. The activities of the design team, professional groups and teaching should be oriented towards such an integrated approach.

Integrated Environmental Design

Integrated Environmental Design (IED) is a philosophy and method of designing buildings which aims to achieve the optimum environmental decisions. Each discipline of specialisation expands its design perspectives so as to be involved in neighbouring disciplines and in the overall design. So, for example, the building services specialists should also be involved in other decisions such as those concerning shape, orientation, use of materials, fenestration, lighting, and sound control.

An integrated approach to design has been found to produce better designs and can give quicker results at no additional cost. An important requirement is that everyone in the design team has a good understanding of building methods, materials, and environment. The performance standards expected of the building need to be defined as early as possible and regular contact maintained with the client throughout the design and construction stages.

Good design intentions for a building can be frustrated by poor installation, inappropriate use and maintenance. There is increasing recognition that the designers of a building should have a commitment to that building for the lifetime of the building.

Useful references

The following publications include sources of authoritative reference and useful further reading:

British Standards, from British Standards Institution.
The Building Regulations, from HMSO.
BRE Digests and *BRE Information Papers*, from Building Research Establishment.
CIBSE Guide, from the Chartered Institution of Building Services Engineers.
CIBSE Code for Interior Lighting, from the Chartered Institution of Building Services Engineers.
CIBSE Window Design, from the Chartered Institution of Building Services Engineers.
Energy Audits and Savings, from the Chartered Institution of Building Services Engineers.
Energy Saving Trust, Internet address: http://www.est.org.uk
Publications from the Lighting Industry Federation

The standards and technology of environmental science are always improving and many aspects of the subjects are of topical interest. Appropriate newspapers, professional and technical journals, information papers and web sites should be used as part of the continuing study of the topics raised in this book.

Index